Scientific Research and Social Goals

Towards a New Development Model

Scientific Research and Social Goals: Towards a New Development Model

Edited by

FEDERICO MAYOR

PERGAMON PRESS
OXFORD · NEW YORK · TORONTO · SYDNEY · PARIS · FRANKFURT

U.K. Pergamon Press Ltd., Headington Hill Hall, Oxford OX3 0BW, England

U.S.A. Pergamon Press Inc., Maxwell House, Fairview Park, Elmsford, New York 10523, U.S.A.

CANADA Pergamon Press Canada Ltd., Suite 104, 150 Consumers Rd., Willowdale, Ontario M2J 1P9, Canada

AUSTRALIA Pergamon Press (Aust.) Pty. Ltd., P.O. Box 544, Potts Point, N.S.W. 2011, Australia

FRANCE Pergamon Press SARL, 24 rue des Ecoles, 75240 Paris, Cedex 05, France

FEDERAL REPUBLIC OF GERMANY Pergamon Press GmbH., 6242 Kronberg-Taunus, Hammerweg 6, Federal Republic of Germany

---

First edition 1982

Library of Congress Cataloging in Publication Data

Scientific research and social goals.
Includes bibliographies.
1. Research—Social aspects—Addresses, essays, lectures.
I. Mayor, Federico.
Q180.55.S62S25 1982 306′.45 82-5214

British Library Cataloguing in Publication Data

Mayor, Federico
Scientific research and social goals.
1. Science—Social aspects
I. Title
306′.4 Q180.55.S62
ISBN 0-08-028118-4

Printed in Great Britain by A. Wheaton & Co. Ltd., Exeter

Acknowledgements

We wish to express our gratitude to UNESCO, whose contribution has made this publication possible, and to Prince Raimondo Torretasso, for his unfailing and generous support of international co-operation and activities for peace.

Contents

Preface: Science, Technology and Social Priorities

FEDERICO MAYOR*

In these introductory remarks I shall not attempt a review of the subject, but try to trace at least its main outlines. In doing so, I will attempt to provide answers—my own answers, naturally—to a series of questions about the role of science and scientists in the development process. Such questions begin by asking why we do scientific research, and for whom its results are intended; what we mean by development, and how its priorities are determined in each country; and how we establish social priorities in accordance with the ethnic, geographical, economic and political characteristics of each country. In all these processes, I believe, science and scientists have an indispensable role to play; they can no longer afford the luxury of withdrawal into their ivory towers.

A short time ago, His Holiness the Pope spoke from UNESCO to the community of scientists, teachers and artists urging them to speak out in order to help achieve a better future for mankind, to ward off the danger of nuclear war, to slow the rush towards conflict and to encourage peace and harmony; and to ensure that the extraordinary sums spent on war or preparation for war—a million dollars every minute—are channelled instead into building a new world in which the dignity of every individual will be safeguarded.

> "Together you have enormous power, [he said] the power of the intellect and the conscience. Show yourself to be more powerful than the most powerful in our contemporary world! Yes, the future of mankind depends on culture! Yes, world peace depends on the primacy of the spirit! Yes, peace in mankind's future depends on love!"

Utopia . . . What is Utopia?

> *S'il est encore temps, il se fait deja tard*—Amadou-Mahtar M'Bow, World Symposium for Science and Culture, UNESCO, Paris, 19 March 1981 ("Although time is still on our side, it is beginning to run out").

---

* Federico Mayor is Minister of Education of Spain, Deputy Director General of UNESCO until 1981.

The Pope's words, like the slogans "solidarity", "new international economic order" or "disarmament", may seem to many a trifle utopian. But I believe that in the strict sense they are not. If they are regarded as objectives that are difficult to attain, and for whose attainment it is necessary to act with an imagination that runs counter to the prevailing current, then they are indeed utopian. But they are feasible utopias, like so many other aspirations and goals which seemed unattainable only a few years ago and which are now taken so much for granted that we do not even think of them. But the fact is that we have very little time to turn these utopias into reality and to achieve it we shall need the collaboration of everybody. We must forget outdated assumptions, shake off the constraints of a crumbling ideology, replace economic objectives by social ones, and advance from political and economic democracy to cultural democracy. It is a process in which everyone must play his part: governments must assume the responsibility that is theirs, and those individuals who have most must sacrifice certain things so that others may acquire the basic necessities. Is this utopian? No; it is simply a question of imagination. Imagination is the most powerful force today, a force that can release us from our present conceptual framework and help us design a new one.

No, this is not utopia. Anyone who still thinks so should read the report of the Brandt Commission, the reports of the World Bank, or the documents prepared for the Third Development Decade. Or they should consider the remarks made by Dr Djomatochona at UNESCO a few months ago, speaking on behalf of the Secretary of the Organization of African States:

> "In our countries, too, which are still suffering from scientific and technological colonialism, inflation and unemployment are giving cause for alarm. Radical changes must therefore be made. We have to acknowledge that current patterns of thinking are inadequate and that nowadays they can be used, at best, only to gauge the extent of the disorder. Our strategy for change is based squarely on a policy of vigorous, bold and constant research. In order to emerge from domination, mimicry and dependence, we must adopt and promote a strategy of endogenous change."

A New Concept of Development

The first questions to be asked about development are simple: What kind of development? Development for whom? In recent decades we have seen that development as conventionally defined, and measured by the enjoyment of material goods does not add up to a greater measure of human happiness even in those few parts of the world which have experienced it—a revelation which has caused much disappointment to those who have (sometimes even without commercial motives) advocated development of this kind. In other countries, this type of development has been resisted by those who perceived in it a threat to their own indentity and dignity, and who saw it as another variant, more subtle but no less dangerous, of colonialism.

Today the only form of development which can command general support is one that emphasizes the well-being of man in the widest sense and that emphasizes welfare rather than wealth. In this type of development full participation by every individual is not only possible but necessary, without sacrifice of personality, identity, or self-esteem.

We are, without a shadow of a doubt, witnessing a rapid shift of emphasis from the economic to the socio-cultural. Our society cannot continue to be based on ideas which are as rooted in the past as those of Adam Smith or Karl Marx. These ideas have shaped the present, but the time has come to discard them. The assumptions are no longer the same; the touchstone is not material wealth but well-being, the total fulfilment of man's potential, and the full exercise of human rights. Indeed, I am sure that the "economic" component will very soon disappear from the organizations which are working to unite countries, or projects intended to lay firmer foundations for the future. Thus in the European Economic Community, only the adjective "economic" is superfluous—when the aspirations are mediocre, the results are necessarily mediocre too. For, ultimately, it should not be an economic community but a European community; it is Europe that is needed today to redress the balance of power and to offer its guidance to the world as a whole. The issue is not one of market adjustment and economic interest—which will always pose difficulties and problems—but of establishing the European *community*.

Why have these changes come about? The great upheaval which began in 1973 with the first oil shock began the change. It shifted the centre of gravity of economic, or at least monetary, power, and of scientific and technological power, which until a few years ago were closely interrelated. In addition, the growth of automation and the migration from country to town has led, first, to a broadening of the range of professional qualifications required and secondly to increasing unemployment at every level. The process has been rapid, allowing no time for adjustment or for imaginative solutions to a problem which is really more apparent than real, since in the long run we should be grateful at the advent of a situation which will enable man to express those abilities which are peculiarly human, leaving machines to do what they do best. The amount of time available for leisure will gradually increase, and we must learn how to make good use of it so that time never hangs heavy on our hands.

These changes imply a major responsibility for social science in solving the problems that beset mankind today and in providing attractive and reasonable alternatives centred on man and not on those pervasive economic considerations which we have allowed to dominate hitherto. In 50 years we may perhaps be working only three or four hours a day, a change which erodes our perception of ourselves and the society we live in. For, as Yves Barel has pointed out, the characteristic which defines Western "modernity" is *work*: not only the products of work, but also the fact that the population at large continues to build its daily life, its psychology, its family life, and its expectations in terms of the "value" represented by work. A man who is out of work thereby feels diminished, not only as a producer, but also as a man. To change this perspective will not be easy, in particular because our knowledge of social systems lags so far behind our understanding of the natural world and of technology. As Aurelio Peccei has put it:

> "The fact that our knowledge and understanding of the natural systems' phenomena and laws have progressed so much more rapidly than our insights into the sphere of societal and human matters in general—which have indeed lagged behind—indicates that a grave cultural distortion affects our proud civilisation. . . . It would be unwise to hide from ourselves that the disparity between techno-scientific and psycho-social development is growing, not diminishing, and hence that, if remedial action is not taken, the human predicament is bound to deepen."

As far as employment is concerned, an entirely original approach is necessary, with forms of work and production geared towards meeting the needs of human beings rather than the market so that a situation may be remedied which would otherwise continue to deteriorate. In this area, it is extremely dangerous for any country to draw up its policies in isolation. With very few exceptions, a country's employment policies should be made known in advance and adjusted to world-wide conditions and trends. For example, it makes little sense to undertake rapid industrialization in certain sectors when a large number of developing countries are doing the same, and market saturation will be reached in a very short time. Similarly, it is not advisable to substitute advanced technology for labour without assessing in good time what the employment consequences of such a policy are likely to be. If policy is to be guided by the criterion of human welfare and by social rather than economic interests, any plan to replace men by machines should be accompanied by a professional retraining programme, designed to ensure continuity of employment, though not necessarily in the same field.

Science and Technology for Development

For most of his history, man has been obliged to put up with nature—with climate, with natural disasters, with the flora and fauna, with epidemics—and has had to struggle for survival with the help of elementary techniques. But today the position, is, as it were, reversed; through the success of modern science, man has gained a command over nature and its resources. After mastering the surface of the earth, he has explored and exploited what lies beneath it; and is now looking forward to the prospect of making use of the ocean, the ocean floor and its sub-strata—an immense reserve of so-far unidentified resources—and the immediate or distant surroundings of the earth, through the control of climate and use of solar energy. The relations between man and nature have thus been completely changed, while instrumentation has become more complex, replacing the direct contact between man and nature reflected in the use of tools and empirical knowledge with the detachment of modern technology. Speculative, abstract science has gradually given way to the applications of sciences, and today what people seem to need most, in order to give purpose to their efforts and guide their future, is reflection on the significance of things: the meaning of life, of human endeavour, of communication and the will to exist. The way, then, that man can reconcile himself to modern scientific knowledge is to bring back theoretical, philosophical, or religious inquiry to the place it deserves; in the same way history and the various human sciences are acquiring a sharper relevance.

In summary, all scientific and technological effort should be directed towards the concept of development outlined, that is development which constitutes more than mere economic growth because it is based on human well-being and not simply on the satisfaction of material requirements, some essential and others less so.

International Priorities

> *Me duele este niño hambriento, como una grandiosa espina*—Miguel Hernandez ("This child's hunger stabs me like a great thorn").

Three years ago, in a work entitled *Prioridades en investigación* ("Priorities in research"), I was able to discuss the major objectives and criteria that influence the choice of priorities. Today, on the threshold of the Third Development Decade, I shall concentrate on the principal factors that together determine the establishment of social priorities.

On a world scale, there are certain basic human needs of unquestionable priority: nutrition (especially during pregnancy and the first few months of life); medical care, universal provision of drinking water; minimum standards of housing; education; and participation in community affairs. These are major problems because they still affect the majority of the world's population; and because the quality of human life depends on the quality of the biological "substrate" on which it rests. Without clean air, water and the provision of adequate food of good quality, human well-being cannot be maintained. The deficiencies of the biological substrate can be illustrated by reference to Bangladesh, where between 70 and 80 per cent of pregnant women are anaemic, and have neither enough to eat or the right kinds of food.

There are other problems which derive from technological development itself. Although the beneficiaries of a development may be relatively few, its ill-effects may be felt by the community at large: pollution, ecological disturbance, the accumulation of carbon dioxide in the atmosphere, or the erosion of the ozone layer by fluorocarbons. Many such problems are made still more urgent because if they go unremedied their effects may be permanent and irreversible. The build-up of carbon dioxide in the atmosphere, from the burning of fossil fuels and the clearance by burning of forests, could have permanent climatic effects. Malnutrition (though not a consequence of technological development) is a similar problem because the effects of too little food on a growing child may well be irreversible. Such *irreversible* problems deserve the highest priority of all.

Scientific and technological research and the transfer of technology must be directed towards the solution of the three problems which constitute the major international concerns of our day: overcoming poverty; the energy problem; and the problem of a more equitable distribution of productive capacity and employment opportunities. Poverty has three particularly urgent aspects: hunger, illiteracy, and disease. Given the rising graph of world population, it is clear that a major effort should be made to increase world agricultural production, and to distribute it more equitably. This is an essential step, without which there can be no new international economic order. Emphasis should be placed on integrated rural development and the rebuilding of a rural society, without which the drift away from rural areas into towns cannot be halted. For the moment we must acknowledge that national health, cultural, educational and economic policies tend to encourage the growth of the cities and rural depopulation; one of the hardest challenges, both at world and local level, is to reduce this tendency, which creates an imbalance which could be one of the most difficult to reverse.

There is no need for me to elaborate on the problems posed by the supply of energy or to describe the upheaval caused by the oil crisis. Two courses of action are equally urgent: the avoidance of waste, and the active search for alternative sources of energy.

Whatever lies in store for us, there are some components that form part of man's essential intellectual equipment: discernment, the possession of basic knowledge, reliance on fundamental principles, the ability to express oneself, the methodological analysis of problems and the capacity to make choices. All these characteristics, which taken together are the hallmark of humanity, can be moulded and enhanced by education and training; but there is no single educational recipe which can be applied without modification to all societies. Recently a group of experts met at UNESCO to discuss how science teaching methods could be adapted to suit the needs of changing societies and the diversity of needs in different countries. They concluded that UNESCO should promote the exchange of information on successful experiments in teaching, and launch studies of new teaching methods.

World Poverty

We must never forget the dimensions of world poverty: one million children in Africa die every year from malaria; 25,000 people die every day because the water they drink is unfit for human consumption; some 800 million suffer from having too little food, and many millions more because the food they have is of poor quality; and it is estimated that last year 30 million children died before reaching the age of 5 because they lacked the basic essentials. Poverty expresses itself in other ways, too. The International Labour Organization recently estimated that more than 55 million children are sent out to work in order to supplement their family income, and this is generally reckoned a conservative estimate.

But although the problems of poverty can be dramatised by aggregrating them in this way, they must be solved by applying the remedies that each one demands. Generalized solutions cannot match the diversity of problems the world presents. A simple example illustrates this truth. Some 200 million people in Latin America, South-East Asia and Africa suffer from endemic goitre. In South-East Asia the cause is iodine deficiency, while in Zaire (where in the north goitre affects 70 per cent of the population) the disease is caused by the anti-thyroidal effect of the local diet based on manioc. In the first case, distribution of iodine supplements would control the condition; in the second, a quite different remedy, reduction in the amount of manioc consumed, is called for.

Each problem demands a specific solution. It also demands continuing effort, even when it seems that the problem may have been conquered. Malaria, for example, a disease spread by the mosquito, seemed to be beaten by the discovery of the insecticide DDT. But the persistence of DDT in the environment caused problems, damaging wildlife and threatening man himself. In addition, indiscriminate use of DDT produced a strain of mosquitos resistant to the insecticide. Worse still, the disease itself began to show signs of resistance to anti-malarial drugs. Today the problem of malaria is far from solved. Like many other problems, it requires continuing, permanent efforts, unceasing vigilance.

Each problem demands its own solution, tailored to the particular circumstances. But at the same time we must not lose sight of the wider perspective. A working group which met at UNESCO in Paris in March

1981 to consider the question of population, resources, and development capacity remarked in its report on the need for the problems of population to be seen in the broader, development context.

> "The relations between population, resources and environment still have to be better understood, demographic factors must be included in development planning and population programmes integrated with social and economic goals. . . . Total numbers of individuals and their distribution, the nature and intensity of agricultural and industrial activity, the effects of this activity on the environment, the material and social aspirations of societies and their cultural backgrounds—all these are intertwined within a complex whole and interventions in one have repercussions, sometimes quite unpredicted, in others. Only by seeing the whole picture rather than a series of partial views can problems be realistically formulated and policies determined accordingly."

National Priorities

With few exceptions, UNESCO's member states establish priorities at national level which are substantially different from those they adopt at international level. Many of the major objectives in UNESCO's regular programme—such as respect for cultural identity, or the framing of science and technology policies—go unmentioned when the member nations of UNESCO are seeking funds in the international market, from the World Bank or through bilateral agreements. In such cases, priority is usually given to things like teacher training, studies on rural settlements, or the training of middle level technicians.

There is no real contradiction here. It reflects, rather, the fact that a country will naturally give priority at national level to goals which are more immediate, better adapted to the political "temper" and therefore more visible to the public than those they support at international level. At national level in a country of moderate resources, it is imperative to choose. Fundamental research is encouraged, but selectively, choosing those fields of research or application which are likely to be of most immediate benefit and which constitute a response to the country's material needs. Such priorities, involving as they do a choice between various options, form a part of government policy and must be examined periodically to see if they need to be adjusted or replaced—for needs change, and national priorities must change too to keep in step with them.

Technology Transfer

The acquisition of technology from abroad poses a dilemma for many developing countries. It offers a rapid and apparently painless way of catching up with the technology of developed countries, but at a price. It must be recognized, for example, that traditional technical assistance programmes have generally reinforced dependence rather than creating independence. The national capabilities of some developing countries have been absorbed into a scientific, technical and economic system which remains dominated by the highly industrialized world, a conclusion which has led many developing countries towards the concept of co-operating with one another, rather than looking towards the industrial world for their know-how.

For what is at issue here is not so much the ownership of technology but rather the gradual acquisition, irrespective of ownership, of the modern "know-how" which is increasingly supplanting traditional knowledge, particularly in agriculture. It is a tendency which gives rise at the same time to a total dependency and to a marked disillusionment among those who believed (perhaps not always wrongly) that inherited knowledge was the best guide to what could be done. The control of this kind of know-how, of "what should be done", is a key source of social power at the present time—as important, though less obvious, than ownership of the means of production.

Technology transfer calls for more than "know-how". It also needs, on the part of the recipient, the ability to absorb and use the knowledge: not only know-how but also "how-know". Without this ability, dependency is more complete and more prolonged, and the development of indigenous capabilities less likely. The nature and organization of higher education is of absolutely central importance. In the past, when change was relatively slow, the rigid "Napoloenic" system with its emphasis on titles and hierarchies worked rather well. The number of students was small, and the tasks they would perform once qualified were well defined. Today the situation is very different; the ability to take decisions is far more important than formal titles. This lesson has been well learned in Japan and the United States, and other developed countries have also introduced a great diversity of new courses, at different levels and in different institutions, in the attempt to train people for an unpredictable future. Many developing countries, unfortunately, are still trapped in the rigid systems bequeathed them as a colonial legacy.

Making Use of Available Knowledge

Today it is a moral imperative to ensure that the mass of people benefit from the application of knowledge already available which at present serves only the few. The acquisition of knowledge presupposes a long and complicated educational process, but its use is a simpler problem. We have a real duty to see that those who are denied access to higher education, those who have no say at the national or international level, are at least able to take comfort from the findings, derive benefit from the discoveries and enjoy the advantages of scientific and technical progress. Inequalities exist not only in the production of knowledge, but also in its applications. For example, thousands of babies are born each year suffering from brain damage which could have been avoided if only a cylinder of oxygen had been available. Equally it has been known for more than a century that chlorination purifies water supplies, but millions still die from polluted water. The outlook for the future may be sombre, but we can certainly change it for the better simply by making better use of the knowledge we have already acquired.

Popularization of Science

As technology advances, the ordinary man's understanding of it recedes. For the untrained observer, there is no worthwhile distinction between advanced technology and magic; both are equally wonderful and bewildering. This growing complexity confers on an élite of technical experts the role of intermediary between the man in the street and the institutions which represent him and which are understood less and less by those who vote for them. It is a role which scientists should not hesitate to play; the scientific community should act as a sounding board, promoting social awareness of the importance and impact of the measures being taken.

The mass media also have an important role. Conventionally they have been spectators, reporting what is going on without accepting a direct role and involvement in shaping the future. This must change, for the people cannot be mobilized without the help of the mass media.

Participation of Youth

Today research and its practical applications are inconceivable without the active participation of youth, without the imaginative and enthusiastic contribution of the rising generation. This is an essential prerequisite for the development of knowledge in any field, but a clear distinction must be drawn between participation in the scientific undertaking and participation in the political undertaking. The right to participate in the former is gained only through the dedication and effort required to gain a good command of the complex web of knowledge while one still has the strength and daring to "think what others have not thought".

Today participation at either level is too limited. We are governed and controlled by a very small proportion of the inhabitants of the planet, and even this ruling minority consists mainly of adults of the male sex. It is impossible to imagine any profound change without the balanced participation of both sexes, and above all without youth, its enthusiasm, inspiration and vitality.

Perspectives and Strategies

Once we have determined our goals, then we are in a position to produce a strategy for reaching those goals. What we cannot do is to devise plans, or fixed programmes, for long-term application, for such rigid plans are inevitably overtaken by events. The objectives should shape the means, never the opposite. What are needed, then, are long-term goals and rolling plans, plans which can be continually tailored to fit changing circumstances. To make this kind of planning work calls for good early warning systems; we cannot foresee everything and to adjust or adapt the plan to events calls for accurate information about what is happening and courage and clear-sightedness from those in charge. The planning process is becoming increasingly complex, not only because of the growth of knowledge but also because of the global nature and impact of the decisions involved. Will this complexity prove too much for man to handle? I am convinced it will not, so long as man creates and the machine assists, and so long as man makes the decisions and does not blindly follow what other men or institutions have decided for him.

The Role of International Organizations

In a world that is dangerously polarized, it is important to have a meeting place for discussion, a role traditionally played by the agencies of the United Nations. But today, although their role is more important than ever, UN organizations suffer from a certain lack of credibility. Some of the reasons are internal: organizations become too big, lose sight of their objectives, and become keener on running the machine than in deciding where it is supposed to be going. There is also "vanity complex", the institutional infighting that goes on within the United Nations as the different specialized agencies fight for their own corner. Other reasons are external: states may damage the credibility of international organizations by tailoring their message too carefully, saying one thing in one place, the opposite in another, for tactical political reasons. In a forum at UNESCO, one state may declare that UNESCO is the home for science in the UN system; in another forum, for a different audience, that a new home for science should be found.

But the UN institutions are even more necessary: they constitute the meeting point of different views and approaches, the place where bridges can be built and consensus reached. They should be more flexible and less hierarchic; but this modern "dinosaur" should not disappear from today's world. The price would be too high.

The trend today is towards confrontation, not interdependence and solidarity, and there are no signs that allow one to suggest any change in the near future. While the disparity between rich and poor widens and while we watch helplessly the decline of the socio-economic systems that have hitherto prevailed, countries both rich and poor are dedicated to arming themselves to the teeth.

We are not talking about the decline of civilizations in the conventional sense; today we are talking about the entire planet. It is a new dimension and a most important challenge. It is not just an empire we are watching in decline, with a limited range of influence and a much slower rate of decay. Today it is a phenomenon that affects the whole world, made one by the methods of communication and destruction.

Governments ought to adopt emergency measures, meet without delay, and decide under the auspices of the United Nations to halt the arms race and then begin the process of disarmament. A book like this cannot in today's climate begin in any other way than by making a resolute appeal that makes people think about the world they live in and that offers the only glimmer of hope; stop the arms race, stop the march towards war! To follow this path would not only be a decision of historic significance for the governments of our day, but also would confer relevance and significance on the United Nations, and would stimulate the younger generation and bring comfort to the whole world. Humanity today is divided between the fight to survive and the fight to understand; it is disheartened and disturbed.

Because we believe that the wisdom of man will prevail, in spite of everything, we have decided to write this book. The fact that it appears at all is a testimony to this conviction and to these hopes.

References

Construire l'avenir, a White Book on research presented to the President of the French Republic (La Documentation Francaise, 1980).

Domination ou Partage?, endogenous development and technology transfer (UNESCO, Paris 1980).

Disturbing the universe by Freeman Dyson (Harper & Row, NY, 1979).

The Global 2000 Report to the President by Gerald O. Barney, Study Director (US Government Printing Office, Washington DC, 1980).

Sciences de la vie et societe by Francois Gros, Francois Jacob and Pierre Royer (Report to the President of the French Republic, La Documentation Francaise, Paris, 1979).

Tiers monde: trois quarts du monde by Maurice Guernier (Report of the Club of Rome, Editorial Dunod, Paris, 1980).

Eighth Plan for Economic and Social Development—1981–1985 (Paris, 1980).

Interfuturs (Informe OECD, 1979).

North–south, a programme for survival, the report of the commission chaired by Willy Brandt (MIT Press, Cambridge, Mass., 1980).

100 pages pour l'avenir by Aurelio Peccei (Ed. Economica, Paris, 1981).

Science and Technology—a five year outlook, the National Research Council (W. H. Friedman and Co., San Francisco, 1979).

Le Defi mondial by Jean-Jacques Servan-Schreiber (Fayard, Paris, 1980).

The state of the planet edited by Alexander King for IFIAS (Pergamon Press, Oxford, 1980).

La Tercera Ola by Alvin Toffler (Plaza y Janes, Barcelona, 1980).

Voix multiples, un seul monde, communication et societe aujourd'hui et demain, the MacBride Commission (Documentation Francaise-Nouvelles Editions Africaines—UNESCO, 1980).

World Development Report, 1980 (The World Bank, Washington DC, 1980).

Foreword

DJERMEN GVISHIANI*

Throughout the history of human civilization the development of science and technology has involved radical changes in human society. It helped solve some problems, created others and transformed the style and conditions of human existence.

As we approach the year 2000, mankind has come up against a number of large-scale, qualitatively new problems which will influence to a significant and probably steadily growing extent the course of world-wide development and the quality of human life.

These problems fall into two major groups. First, there is the whole range of issues related to the reshaping of international relations, including prevention of nuclear war, scaling down of the arms race, further development of international co-operation based on peaceful coexistence of countries with different social systems and the restructuring of the world economic order. Second, there is the complex relationship between man and nature, which includes the satisfaction of the growing demand of the planet's increasing population, taking into account the ever greater human impact on the natural habitat.

Nowadays, when contacts between peoples and countries have grown in extent and diversity, giving rise to numerous areas of contradiction and opportunities for international tension, the normalization of international relations—putting an end to the arms race and the transition to actual disarmament—has become the most important global problem and at the same time an indispensable prerequisite for solving all other global problems. Without it there is no sense in speaking about international economic relations on a basis of economic equality and mutually beneficial co-operation.

The backwardness, dependence and poverty of the majority of the developing countries have their roots in those countries' colonial past. Recently the general situation in most of the countries of Africa, Asia and Latin America has got much worse, and their two "Development Decades" have in practice turned out to be decades of disappointment. A small quantitative rise in individual indicators of economic and of scientific and

* Djermen Gvishiani is Deputy Chairman, USSR State Committee for Science and Technology and Director of the Institute for Systems Studies.

technological development cannot hide the fact that, in qualitative terms, the developing countries are falling further and further behind the developed countries and have to contend simultaneously with an enormous rise in external indebtedness, a severe capital shortage, a colossal labour surplus leading to mass unemployment, the relative contraction of commodity markets and so on.

A social structure which will make it easier to assimilate contemporary scientific and technical progress must be considered a decisive precondition for development. This is, in its turn, linked to active participation in economic, scientific and technological development.

That is all the more so because the development of science and technology is an important factor (though by no means the only one) in scientific and technical progress. Success depends not only on a country's level of research and development, but also on the conditions under which the results of that research are applied, conditions determined by the structure and level of production. Fruitful application of the results of the scientific and technological revolution requires from the developing countries a rapid switch to the use of intensive economic methods, since the extensive approach, while it may ensure the rise of quantitative indicators, will be marked by a progressive slowing down in respect of other parameters—quality, growth rates and levels of development. In other words, the tempo of capital formation can be accelerated only by combining the industrial with the scientific and technological revolution.

The impact of science and technology on society has grown immensely. Humanity has never been so thoroughly equipped for solving the problems that confront it. New discoveries in physics can help create new energy sources, based on thermonuclear fusion, although it will take more than ten years to solve the purely technological difficulties. Great results are expected in the computer field (micro-mini-computers, memory, etc.) and development of new superstrong materials. Insight into the composition and structure of molecules, development of kinetics in chemistry, new principles of catalysis and control of chemical reactions, as well as many other lines of research open up new prospects for creating substances with useful properties (plastic materials, semiconductors, magnetic and superconducting materials). In this way chemistry can make a tangible contribution towards the rational use of natural resources. Modern achievements in biology are of great importance to the solution of food and ecological problems. Intensive research at the borderline between two or more sciences (biochemistry, geophysics, biophysics, etc.) can yield very significant but so far unpredicted results. Science has proved helpful in solving many difficult problems.

At the same time the progress of science and technology can bring about some rather complicated problems. It is not, in fact, the progress of science and technology itself which is responsible for these disappointments and mistakes, but some of its social aspects and implications. Many scientific discoveries and technological inventions were and are being made within the private entrepreneurial system, where their impact on society, nature, and so on, is often ignored. Because cheap raw materials and fuels were available scientific and technological progress took a power-intensive and resource-intensive route which, in its turn, added to the energy and raw material crisis when it arrived. Big profits were made by ignoring environmental standards and some countries are paying a high price today for that negligence. The negative side of modern scientific-technological development has created in some circles the view that science is antihumanistic and that we should consider stopping or limiting its development. I am convinced that science, as an important part of our common cultural heritage and as a powerful product of human activity, can and shall contribute to satisfying human needs and aspirations.

To bring socio-economic development into harmony with people's aspirations for peace, freedom, work, participation, and security, science should be compatible with socio-human values, and a science policy should be oriented towards broader social goals. What we really need is a new approach to development, or a new development model. Development will be taking place, but on a qualitatively new basis; and this is why, when we talk about the development of science and technology, it is extremely important to understand the consequences of the decisions that we are making today. I think major changes are going to take place in every country because we are functioning, or beginning to function, under new conditions. More scientifically-based policies have to be established to provide guidelines for the development of society. Rational management, based upon comprehensive study, is an absolute necessity, and it is extremely important that there should be international co-operation on the issues that are important for humanity.

Although the prospects for co-operation are not good at the moment, I nevertheless strongly believe that goodwill and co-operation, combined with a frank acknowledgement of the differences that exist, will help to solve our problems. One cannot expect the solution of all these problems at once; that would be unrealistic. What is particularly important today is to cherish those cases where agreement *is* possible, whether this is in the scientific, the technological, the economic, the cultural, or the political arena. I strongly believe in the wisdom of people; people who will in the end be convinced that they have a joint responsibility for the survival of humanity.

We live—I am being realistic—in a world that has different values and different orientations, and therefore I do not think that it will be simple to work out standard solutions or patterns which can be strictly followed throughout the whole world. There might be, in such standards, a bias which reflected a specific society or a specific socio-economic system. At the same time, some parts of these standards really will be universal.

Global problems make the world aware of the need for international co-operation, including co-operation in science and technology. By its very nature, science is well equipped for internationally co-ordinated efforts directed to the solution of common problems. Science is universal, independent of nationality, ideological convictions or political orientation, which makes joint efforts much easier than in any other field. Scientists can assess facts rationally and settle disagreements on scientific matters. And if science is to have an influence on global problems, the financial, human and material resources needed are so great that they can only be put together through international co-operation.

The large-scale, comprehensive and multidisciplinary nature of all these problems requires the application of a systems approach. Many of the important tools (concept standardization methods, unified analytical methods and procedures, formalization of scientific language, interpretation of results, and so on) require thorough discussion and the use of all the expertise available in various countries. The Soviet Union has always attached great importance to the development of international programmes of scientific and technical co-operation in this field.

There will be a vast variety of different problems, each calling for a different form of co-operation. Of primary importance is the expansion of data collection and information exchange, as well as their analysis and generalization, and subsequent discussion of basic conclusions. To this end, we should develop wider scientific contacts and links on a bilateral, regional and global basis. In many spheres there already exist the structures necessary for co-ordinating research plans and formulating international research programmes. Some thought should be given to the idea of setting up international research and, later perhaps, research-production complexes. The experience of institutions such as the International Institute for Applied Systems Analysis (IIASA) and the CERN Institute in Geneva shows how successful scientists from different countries can be when co-operating to solve the complex problems of our time.

The choices we make today in science affect the future in ways difficult or even impossible to envisage. Governments are very often guided by immediate objectives and national policies in science and technology more or less reflect these immediate concerns. To reach overall conclusions, it is extremely important to have a sort of independent assessment, especially on issues which require forecasting. This is the justification for the organizations already devoted to this matter, and perhaps for the creation of some new ones.

In this connection I would like to mention the International Institute for Applied Systems Analysis in Vienna. IIASA, a nongovernmental, multidisciplinary, international research institution, was founded in October 1972 by the academies of science and equivalent scientific organizations of twelve nations from both East and West. Its goal is to bring together scientists from around the world to work on problems of common interest, particularly those resulting from scientific and technological development. The institute now has seventeen national member organizations. The studies carried out in the institute attempt to view global and universal problems in the broadest possible terms. What I think is particularly important is that it helps to create a sort of independent expertise, a scientific assessment independent of national governments.

A prominent role in the solution of the crucial problems facing the contemporary world has been given to the United Nations Organization and its specialized agencies. It can act as an organizer and co-ordinator of a wide variety of scientific and technological studies, making use of the member countries' expertise. For example, the question of the reorientation of scientific research was touched on at the UN Conference on Science and Technology for Development held in 1979 in Vienna. The preparatory work for this important undertaking included symposia in Tallin (USSR) and in Singapore. A group of Soviet scientists made an assessment in their respective fields of science, and the results were presented at the Tallin meeting, the International Symposium on Trends and Perspectives in Development of Science and Technology and their Impact on the Solution of Contemporary Global Problems, held between the 8th and the 12th of January 1979. These surveys are evidence that some sciences are already making a contribution to the solution of the problems under investigation.

The task the world faces is to humanize the development of science, to direct scientific research towards the satisfaction of human needs, and to promote among scientists international co-operation on a humanistic basis. As the international agency with the mandate to promote the development and integration in the fields of culture, science and education, UNESCO has always paid attention to the humanistic aspects of scientific research. Its programme on research and human needs* is intended to throw light on this very complicated issue.

The contribution of science to the solution of almost any social problem can be made in several alternative ways. The choice of alternative is determined by the social and cultural development of any particular country. It is important, therefore, that in creating a science policy one should take into account its general social and cultural context. The determination of the relationship between research and social needs is faced with uncertainty; it demands from scientists not only high scientific expertise and competence, but also a vision of contemporary human problems and an acute sense of

* Contribution to the determination of research priorities linked with human needs and societal goals.

responsibility. Scientists must also develop adequate methodologies so as to foresee and forecast trends in scientific research and its future potential, which is becoming more and more important in our rapidly changing world. Achievement of these goals presupposes close co-operation among scientists, representing different scientific disciplines, different countries, and different approaches, cultures and ideologies. I am absolutely convinced that it is only through this co-operation that modern science can meet the challenge of our time. In this sense the book is not only an endowment to the humanization of research, but also to the development of the co-operative spirit within the scientific community. Contributing to a better mutual understanding, meetings and discussions within the framework of the programme are one step—perhaps a small one—towards a really international science of the future.

Particularly important to me is the fact that the programme represents different scientific disciplines and countries and reflects various methodological approaches and ideologies which coexist in our complex world, without trying to force a reconciliation which would be acceptable to all the members of the international group of scientists. This false kind of unanimity, especially in science, is not only unnatural but also very often leads to the oversimplification of real problems.

There are, however, points upon which all the authors of the book and all the scientists dealing with this complex problem seem to be absolutely unanimous. These are the necessity for joint efforts of the world scientific community and for close international co-operation.

On behalf of my Soviet colleagues and of members of the Institute for Systems Studies and other institutions who participated in this programe, I would like to extend our gratitude to Deputy Director-General of UNESCO, Dr Federico Mayor, whose intelligent approach and personal attention to the issue has been a constant influence. We also express our gratitude to Dr Augusto Forti for his constant assistance and helpful co-ordination, as well as to all the members of the international team for valuable discussions and co-operation.

Introduction

AUGUSTO FORTI, HENRI HOGBE-NLEND, OSCAR NUDLER AND OLGA ZAKHAROVA*

Why a New Approach to Development?

We are living at a time when new discoveries in science and their technological applications offer more and more opportunities for rapid socio-economic development. At the same time, the directions and goals of this development are being questioned more and more all over the world. Instead of contributing to human fulfilment and the improvement of life, development appears to cause a lot of new problems without solving the old ones efficiently.

Over the last few decades the notion of development has become common currency in political debates as well as in discussions within the scientific community. But the concept has different meanings depending on its context and on the ideological and political convictions of those who use it. Another popular idea is that "progress" in science and technology is supposed to contribute to social development. But it becomes more and more evident not only that the aims of development do not always correspond with the necessity of finding a solution to crucial contemporary problems, but also that the idea of development itself needs a new approach, which instead of reducing it to economic and technological progress would focus it on real human needs and problems facing humanity as a whole.

Doubts arise as to whether the economic indicators of development really correspond to human aspirations for a better life, or fuller satisfaction of their material and social needs.

The most widely understood meaning of the word development would include such elements as the growth of GNP, quantity of goods and services, and so on. But the growth of these material benefits does not necessarily lead to their more equal distribution and also makes no contribution to the satisfaction of spiritual needs, which are no less important for human beings than material ones. Alienation from society, growing frustration among young people and other negative processes in economically and technically

*Augusto Forti is in the Division of Scientific Research and Higher Education, UNESCO; Henri Hogbe-Nlend is President of the Pan-African Mathematical Union; Oscar Nudler is at the Bariloche Foundation, Argentina; and Olga Zakharova is at the Institute for System Studies in Moscow.

developed countries are important indicators that in these countries the situation is far from satisfactory.

The need for a new approach to development is now widely acknowledged. The clamour for change comes from everywhere: from scientists, scholars, statesmen, and above all from the powerless and innumerable poor. Today's political and economic system appears to them to have failed; and as a result of this failure, the international organizations are now all engaged in the attempt to create a New International Economic Order. The world is in transition; an uneasy state of mind which is produced when one faith has been abandoned but no new one has arisen to take its place. That is why to talk of the search for a new development model is no mere intellectual abstraction.

Nor will the new model be produced by intellectuals alone. It will evolve out of a social process in which everybody will have to participate. Each nation and each social group has the right to build its own future on its own social and cultural heritage, but today the interdependence of different countries and different regions must also be considered. The challenges to be faced in the year 2000 will also be, in great part, global in scale. To reflect this global interdependence it is essential that countries should develop in harmony with one another.

In this process of discovery, intellectuals have both the right and the duty to participate. History has shown more than once that ideas can have a profound effect on the way men behave, and that thinkers can open up alternatives that have never been considered seriously before. This book is designed to contribute to the task of producing a new development model, based on human needs, social goals, and the value of research.

Why Human Needs?

One criticism almost universally voiced about today's development model is that it is crudely materialistic, its success or failure measured by the criterion of economic growth. In the developed countries, this philosophy is showing its drawbacks more and more clearly, in such things as stress, pollution, alienation, unemployment and inflation. In the developing countries, criticism centres more on the fact that the "trickle-down" process has so far developed only small élites, leaving most of the population untouched or in worse condition than before.

The question then is: what should be the new development objectives? Since we believe that development should be centred on the equitable and sustainable development of all human beings, examination of the process should start by looking at what people's needs—personal, social, and ecological—really are. Only in this way can a valid international development model be established.

Why Societal Goals?

While human beings can shape their own lives, it is clear that they are strongly conditioned by the society in which they live. Their chances of self-fulfilment are determined very largely by the structural characteristics of that society. This means, of course, that the goals of a society and those of the individuals within it are very closely interwoven. In ideal circumstances the individual by pursuing his goals helps society to achieve its goals, and vice versa. This is why it is irrelevant to speak of the satisfaction of human needs without relating them to social structures, or to talk of society's aims as if they were distinct from those of the individuals making up that society. It is the individual, not the society, which should be the object for development. But to understand the process of achieving that development calls for analysis from many different fields of knowledge, as well as an appreciation of the local and cultural peculiarities.

Why Research?

Science has given us an understanding of the world, classified by discipline, but little real comprehension of how these disciplinary perspectives can be assembled to provide a complete picture. (This shows both the strengths and the weaknesses of the Cartesian approach to understanding which we have been following.) The same is true for social problems; we know how some of them can be tackled, but we are defeated by the larger task of solving several at once, particularly when they transcend narrow sectoral definitions.

The effective response of science to these challenges depends first on how far scientists perceive their social responsibilities, and secondly how far they are prepared to transform that perception into relevant activity. They will need both scientific knowledge and understanding and a deep appreciation of society's goals. In the immediate future the greatest problem is not the accumulation of more specialized knowledge, but the concentration on integrative and "holistic" knowledge which transcends disciplinary boundaries. The challenge is a daunting one, but society needs answers if it is to solve its problems and improve the prospects of the human race. It is not simply something which a few intellectuals have dreamed up in their ivory towers; quite possibly the survival of the human species and certainly the survival of much of humanity depends on its solution.

The Contents of the Book

This book contains the most important results of the UNESCO programme on "Research and Human Needs".

The 1970s has been a decade of anticipation and alarm. Warnings about the precarious situation of mankind have been made by many people, as well as by the international organizations of the United Nations, and by non-governmental organizations. The 1980s must become the decade of political will and action if we are going to see the changes which we know to be necessary. One of the most important lessons of the 1970s is that "global problems do not necessarily imply global solutions". While the interdependence of all humans and all ecosystems makes it imperative to think globally, in practice each social group has to solve its own problems, even if outside help is provided. We must not allow the vision of some future Utopia to blind us to the claims of those who are suffering here and now. The millions of children being born into the world every week to face the permanent injustice caused by malnutrition and poor health care are not helped by promises of some distant paradise.

We seem trapped, however, with old concepts and worn-out remedies, still hoping that these will function as they did in the past and that we shall soon be back to "business as usual". For more than a quarter of a century, the emphasis of development has been economic growth; the goals of the two UN development decades have been expressed in these terms, and the well-being of society has been measured by economic indicators. We now recognize that this is not enough. Development goals which take the social and cultural dimension into account, and which cater to the entire population—not just a privileged section—are urgently needed. We need a holistic approach to establishing society's goals, which will enable us better to understand the problems, and will help us to mobilize our physical and mental resources to solve local problems in a global perspective. For this, new development models are needed, and our research efforts must be directed towards real problems and genuine human needs. There are several examples in recent history to show that scientists can be mobilized successfully to achieve politically defined goals—for example, the wartime atomic bomb and radar programmes and, more recently, the Man-on-the-Moon programme.

The "Research and Human Needs" programme was conceived at a symposium/workshop in Venice in December 1975. A careful plan was worked out, covering the scope, purpose and implementation of the programme, and in 1976 a number of reports were published which helped to bring the problem to the world's attention. This book discusses the results so far achieved, and what they imply for the future. It is divided into three sections: concepts and philosophy, methodological approaches, and practical programmes in different parts of the world.

We hope that the encouraging involvement of scientists, scholars and policy-makers in this programme will be only the beginning of a world-wide mobilization of intellectual and physical resources in joint efforts focused on human needs, rather than on increasing the destructive capacity of our small world.

SECTION 1

Research and Human Needs

Research and Human Needs: a Challenge and a Hope

ANDRÉ DANZIN*

Scientific research reflects a very special and fundamental need of mankind: the need to understand, to explain, and to know. The search for knowledge has always been an inspiration. By its very nature, research explores what is unknown; the scientist is always prepared for the unexpected, and he knows that what he is going to discover when his work is complete could be very different from what he expected, could even turn out to be the exact opposite of the hypothesis on which he based his experiments. He is willing to accept the most far-reaching changes in the reality created by his mental models. This willingness to embrace the unexpected has always been emphasized by scholars; the message was expressed particularly well by Claude Bernard in his *Introduction to Experimental Medicine,* published in 1865.

Science itself is characterized by logical rigour, but its effects on human society often appear irrational. What one finds is not necessarily what one expected, and the results are, as often as not, much more significant than the "expected" result which originally inspired the work. If one ignores the unpredictable and irrational character of research, and if one limits the freedom of the research worker by giving him a specific objective, one strikes a mortal blow at the whole process of scientific productivity.

But the popular view is rather different; it sees the application of science as a perfectly rational process. Man lands on the Moon, explores space, sails to the farthest corners of the world, eliminates diseases, transplants organs, communicates over huge distances, travels faster than sound, controls nuclear energy. Man becomes all powerful, thanks to the great god science. "If science were to be properly directed, if it could be freed from pressures that pervert it—the arms race, and the constant search for profit—it would provide all the answers man needs to solve his problems." This is a view one often hears expressed by people who have no idea how scientific research works.

---

* André Danzin is director of Research at the Institut National de Recherche d'Informatique et d'Automatique (INRIA), and President of the Association Francaise de Cybernétique Économique et Technique (AFCET). The author is indebted to Professor I. Prigogine, Nobel Prize, of the Université libre de Bruxelles, with whom this article was discussed extensively and who shares the views of the author.

Given the obvious and acute need to find answers to these problems, there is something rather sad in this dilemma: how, when one knows that the very essence of science is its uncertainty, can one be sure of satisfying a hope based on a mistaken notion of what science is?

Difficulties increase as one begins to analyse what human needs really are. Man's needs are in fact extraordinarily varied, depending on where he lives and the stage of development of the society he lives in. The situation is so complex that one cannot give a simple description of what it is that mankind really needs. Even the perception of needs changes. Man is a creature more of desires than of needs; and as soon as his desires are satisfied, he tends to ask for more. It is in his nature to do so. Man is made for evolution, a process which constantly turns up new questions, some of them born from the very answers to the previous series of questions.

So it is not only the complexity of the subject, and its huge regional variations, which prevents one from putting forward a few models which would serve as ideal examples of development. The real reason is that what seems sensible to us today would probably be useless in a little while. How does one create a policy for science in a climate which is so hard to define and which is in a constant state of change?

But this analysis may seem to offend against common sense. Science has demonstrated its power most convincingly in the development of new technologies. During the past hundred years it has been the most powerful cause of change in the economy, in morals, and in ideas. It is difficult to admit that we cannot direct this force and put it to the service of satisfying human needs—some of which, like malnutrition, the energy crisis, and the conquest of illiteracy, are clearly of the highest priority.

Evolution at the Crossroads

Before going any further and risking giving a false notion of things, we must stop to think for a moment about the nature of change. When a phenomenon is as complex and as subject to alteration as human society undoubtedly is, its evolutionary path consists of long periods of continuity interrupted by abrupt changes of direction (Figure 1).

If we study a particular phenomenon—economic activity, for example—we discover that it begins by following a smooth curve (section 1). Let us assume that it grows exponentially, supported by a constantly growing consumption of cheap energy, until there comes a moment when these sources of energy show signs of running out. One reaches a crossroads (B) where a choice of new directions is offered, none of them identical to that of the curve before. For example, one might be able to discover no way of solving the energy shortage, and economic activity will decline (curve 2b); alternatively, by making use of new sources of energy, and recycling materials which are in short supply, economic activity may continue to increase, though at a slower rate (curve 2a). A third possibility is that the economy changes its whole nature, turning towards the consumption of new materials which feed a different kind of economic growth following different rules (curve 2c).

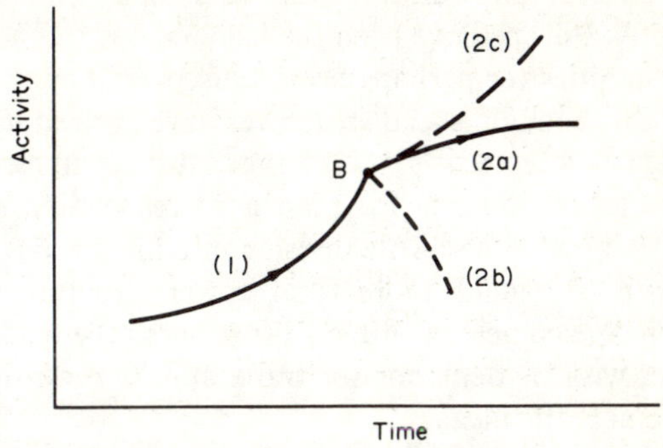

Fig. 1. Discontinuous evolution

If we look at physical or biological phenomena, such as the behaviour of water molecules boiling in a test-tube, or a colony of micro-organisms in a culture, the arrangements can develop and change almost spontaneously. When a moment of change is reached, where the curves divide and many different courses are possible, it needs only the slightest intervention, or "trigger", to send the system one way rather than the other. But once this change of direction has taken place it is irreversible and the system develops in its own way until it reaches a new point of instability.

In human societies the social environment, or "milieu", is created by patterns of consumption, structures of change, morals and ideas. Some eras are seen as periods of continuity, even if they led up to profound changes; the ruling class of Europe in the year 1900 looked forward to a 3 per cent return on their investments for ever, and no change in the structures of power. The last quarter of the 20th century, by contrast, seems to us to be heading for a moment of change, even if we are incapable of guessing how things will evolve after that change. The developed world lives in fear of economic crisis, inflation, unemployment and energy shortage; the oil-producing countries have established an illusory wealth which they do not know how to invest and of whose future they have no guarantee. In China attempts are made to improve the individual standard of living while each year brings 10 million more mouths to feed and 10 million young workers in search of employment. The situation in the poorest countries is even more alarming.

Science cannot produce answers to problems whose nature is more moral and political than technical. But it can have an indirect influence: first by modifying the social environment, second by acting as a "trigger" for change.

Changing the Environment

From the very earliest times, technology has always modified the social environment. Neolithic societies of the Bronze Age, the Iron Age and the other successive waves of human evolution were not established simultaneously all over the world. At different times China, Egypt, Greece and, more recently, Europe have been the leaders, but technology has spread to isolated communities (or perhaps been rediscovered by them) much more slowly. Remarkably enough, social structures have evolved in much the same way in each region affected by the change, just as if for every stage of technological evolution there is for man a corresponding social structure. Neolithic society probably had little division into hierarchies, but the Bronze Age brought with it a much more rigid power structure, often based on slavery. The Iron Age coincided with a shift towards the feudal system, while industrialization, with its demands for trade and its rewards for enterprise, produced new democratic ideas.

Of course, the social environment is not changed by technology alone. It is the interaction between philosophical ideas and the applications of new technology which produces the change. Which comes first? This is really no more interesting than asking whether the chicken or the egg came first, because what we are considering is a continuous process of action and reaction. The more social conditions and the political climate favour the emergence of new technology, the more is produced. At the same time, technical progress reinforces the existing social structure, at least until a certain threshold is reached, or success becomes gross and contradictions emerge.

In fact it is the growth of knowledge which is principally responsible for modifying the social fabric. Bit by bit, man accumulates knowledge and information; from this he derives a growing power over nature and an increased ability for self-examination. The ascendancy over nature appears to be limitless; it creates what one might call a "rule by machines". And human introspection is limited by the problems of complexity, which increase just as rapidly as knowledge does. Because of the growing interdependence of the different parts of the economic and social system, no part can be analysed on its own, detached from the rest.

In all the eras which preceded the scientific age, knowledge advanced in a loosely organized way; small groups of men gathered to form a school of thought, or forceful individuals dedicated themselves to scholarship and invention.

Some periods are notable for the progress made in acquiring technical skills; others in the creation of new philosophical and political ideas; yet others seem anchored in conservatism, wedded to certain ideas which have proved successful in the past and still enjoy ideological support. At any given moment, all these different attitudes can be seen in operation somewhere in the world.

Nevertheless, our period is characterized by a new phenomenon, created by the invasion of science and technology into the upper levels of our society. This does not differ in kind, but in degree, from what happened in the past. Certainly Pythagoras, Pascal, Leibniz and Newton played major parts in the discussion of ideas in their own time; and they had no hesitation in being philosophers as well as scientists, to the extent that it is sometimes hard to decide which was more important in their work, science or philosophy or the cross-fertilization between them. But these were exceptional individuals. Society then had not conceived of the idea of mobilizing the teams of scientists we see today; a transformation which calls to mind the movement after the French Revolution from military élites and mercenaries to popular mobilization of the whole male population. Since the end of the 1950s those who work in laboratories in the developed world can be counted in hundreds of thousands, or in millions if we include technicians and administrators as well as scientists. The consequence has been an explosion of new knowledge, with profound implications for the social and economic system.

Curiously enough, at the very moment scientists were acquiring *de facto* their position as principal agents of change, they were attempting to retreat into the isolation of their specialities, renouncing their philosophical background. Instead of acknowledging the global nature of knowledge, most scientists were trying to restrict their efforts to narrow technical questions under the pretext that only in this way could they preserve their objectivity.

There are already signs that in the years to come we are going to see an end to this isolation. Scientists want to throw off the narrow rules which have led them beyond objectivity into what might be called "objectivism"; both conscience and an awareness of the effects their work may have will force them to renew ties with society. In this way a "new alliance" will be born, transforming science in a way which we can already begin to see. Even if it is not yet clear how it can come about, one can hope that the separation between natural sciences and human sciences will gradually disappear. The transformation of science should be marked by a renewed ability for scientists to move relatively easily between one speciality and another, and by a closer involvement between science and human values. The message of science will then cease to be purely technical, and it will be possible to understand man in all his complexity, in his need for social stability just as much as in his desire for personal freedom.

A science policy directed along these new avenues could act as a powerful force for change, and most probably a wholly beneficial one. Properly applied, it could help man understand the full range of problems he faces today and will face tomorrow. It could help to distinguish between real human needs, and the false gods created by ideology. The alliance between science and society could thus help to prepare the ground for a new culture, and provide a new spirit so often sought by today's social critics. Everything possible must be done to help reconcile the natural sciences with what man perceives as his human rights, and how this might be done is beginning to emerge: human mobility, the transfer of technology, the *rapprochement* between human and physical sciences, the growth of information technology and the reform of education could all have a part to play.

Science as a Force for Change

But as we have already seen, preparing the ground is not enough. It is also necessary to envisage a decisive event—in our earlier analogy, a "trigger"—which opens the way towards one solution among the many which seem at present to be equally possible. These triggers are always available; often they fail to ignite the charge, but it is important that they should sometimes succeed. Triggering events may often pass unnoticed by most observers; and if they are noticed, their importance is often underrated, or even their existence denied, by those who are closely linked with the existing power structure and who have the most to lose by change.

The remote past can provide examples of such agents of change. The discovery of fire, the invention of language, the use of the wheel, the creation of shape—first with ceramics, then with metals—then, closer to our own time, the development of harnesses for draft animals, the invention of gunpowder, printing and the magnetic compass all seemed like accidents to which very little attention was paid at first but which later turned out to be powerful agents of change in human society. Pasteur could not have guessed that his work with micro-organisms and protection from disease by vaccination would lead ultimately to a new attitude towards life and death because of its dramatic effect on the growth of populations. The engineers responsible for developing the various kinds of heat engine had no idea that they were helping bring into being a civilization which would conquer air and space. We ourselves are living through a revolution in information technology which is transforming the way knowledge is stored and exchanged, touching the very core of our culture.

There is not much chance than man will ever be able to predict the long-term consequences of such changes, but at least he should by now be aware that such processes do effect human evolution. He can then attempt to control them, either by strangling them at birth if their effects are by general agreement likely to be disastrous, or by encouraging them if their results seem likely to respond to a real need. This process of planning by looking out for key triggering events calls for some thought. In the state of uncertainty we find ourselves in today, the only certainty we have is that things cannot go on the way they are, and we must accept that an abrupt change lies ahead. If so, then everybody must be allowed to express their point of view. One can bring "imagination to power" by allowing inventive ideas to come forward, and one can also select those specialist areas where new ideas are likely to be most productive. For example, it is probable that tropical agriculture might be transformed if one could attach nitrogen-fixing bacteria to the roots of cereal crops; there is thus every reason for a special effort in this area. The process then becomes a question of political priorities, but one must always keep an eye open for the unexpected. The discovery of penicillin, partly a lucky chance, turned out to be much more important than all other contemporary work on chemo-therapy.

Science and Know-how

In practice, as Bertrand de Jouvenal has remarked, our technological society has always been much more interested in "doing" than in "knowing". What we have been looking for is not so much increased knowledge, but an increased power over nature. This preoccupation explains why scientific research is often confused with what might better be called "know-how". It is a disastrous confusion. For one thing, it conceals the role research can have in changing the social environment, or in providing a supply of "triggers" for such change. For another, it encourages the illusion that all would be well with the world if know-how could be more easily spread and hence more equally shared. If technology can get us to the Moon, surely it can solve the problems of under-development!

Without denying that progress could be made if our society were better educated in science and techology, it would be naïve to believe that the huge questions facing humanity today could be so simply solved. The technology we have is relevant only in particular places, at particular times, and under particular political conditions. Western society is not exportable; it consumes too much energy and too many primary products. High levels of mechanization are not suited to the growing pool of unemployed manpower in the developing countries. Agricultural techniques work only in a temperate climate where the land is good, exposed neither to drought nor to soil erosion. They do not produce a high return without the use of fertilizer and pesticides, and they consume through the use of machines many more fuel calories than are produced by the food which is grown. Such a solution cannot be applied generally in any situation, nor is it likely to survive for ever, even in the developed countries.

One can multiply examples: if improved health standards prolong life and reduce perinatal mortality (as they do) and if they are not accompanied by a reduction in birth-rates, then population problems sooner or later become insoluble; which brings the matter back into the realm of morals. To believe that all that is necessary is to extract the knowledge one needs from a reservoir, like water from a well, is an enormous error, unfortunately widely shared. On the contrary, what is needed is to establish basic research facilities and so direct them that they follow useful paths leading to the solution of the problems of underdevelopment. The science of the North has until now taken little account of the needs of the South, and the research workers of the South, often excellent in quality if limited in number, have not concentrated on problems relevant to their own areas, absorbed as they are in the science of the developed world. It is time for a change.

Confidence and modesty ought to be the key-words in the dialogue between scientists and politicians looking for a better world system and a reduction in poverty. Confidence in research as a means of understanding the situation, so long as one is prepared to look at problems in their full range. Modesty as to immediate results: everything that has been said goes to show that the road will be long and sometimes bumpy and that when it comes to meeting human needs, science can do little without political will.

*National Policy Implications of the Basic Needs Model**

K. SOEDJATMOKO†

We all know that the basic human needs approach to development grew out of the search for a development strategy which could deal more effectively with the problem of continuing poverty in large parts of the world. Such a strategy should, within the limited timespan of 25–30 years, be able to reverse the trend of growing inequality in developing societies, which threatens permanently to leave the poorest 20 per cent of the world population in conditions of absolute poverty. The basic needs approach constitutes a direct attack on world poverty by meeting the basic needs in the fields of food, nutrition, health, education and housing, as well as through employment and income-generating activities among the lowest 40 per cent income groups.

It is predicated on a policy package consisting of a relatively high growth rate (6–8 per cent), redistribution of income, and—up to a point—wealth, reorientation of investment, and a review of consumption and production patterns. It is hoped that such a thrust, with the rural sector and labour as major factors in productivity increase, will lead to processes of internally generated development, which over time might develop into an indigenous engine of growth.

No wonder that the basic needs model has, in a relatively short time, gained such popularity, and has even become fashionable in many quarters of the international community. In a way deservedly so. The basic needs model is, undoubtedly, an important contribution to development theory, of great moral appeal. It provides a promising point of entry into a development problem, which seemed beyond the effective reach of earlier strategies. It has, in fact, already led in various countries to important adjustments of existing development programmes. Nevertheless, it is also true that while important

---

* This paper is a slightly revised and shortened version of one presented at a seminar on "Implications of the Basic Needs Model", organized by the Dutch National Advisory Council for Development Co-operation, held at The Hague on 24 February 1978. The author wishes to express his thanks to Sayuti Hasibuan, Stijn Albregts and James Grant for their valuable comments on an earlier draft of this paper. The author remains, of course, responsible for any mistakes or errors of judgement in the paper.

† K. Soedjatmoko is the Rector of the UN University.

work has been done in developing data in support of the concept,[1]* as well as in developing policy packages in the priority areas which the basic needs model is concerned with,[2] surprisingly little has been done in studying the national policy framework for development which the basic needs model requires for effective implementation; its relationship to other national development goals, or to the economic-political dynamic of the development process itself. Several studies mention the need for "political will" as a prerequisite for the implementation of the model, but all remain silent on how to generate such a will.[3] Nor do they consider the political risks and costs involved, the difficult economic and political trade-offs that have to be made, and other perplexing dilemmas that have to be faced.

This paper is a modest, tentative and necessarily subjective effort to look at some of the broader domestic, as well as international, implications of the basic needs model for national policy. It raises the question of the capacity of political systems to absorb the requirements of the model, of the nature of social change and of the development process which constitutes the environment within which the basic needs model has to be implemented, and of the ideological presuppositions which bear on the choice of methods in meeting basic needs.

One point should be made at the outset. Although it is often presented in a very programmatic fashion, the basic needs model takes us into a number of areas about which, collectively, we really know very little. We still do not know, for instance, despite a few obvious success stories, how to bring about rural development. We know little about how to stimulate small business enterprises in rural areas or in urban informal sectors. We do not know very much about the income—and employment—generating linkages in the rural areas, and even less about how these are affected, directly or through different technology choices, by fluctuations in food prices. While we know something about how food is produced, consumed and imported on a national scale, we know very little about food consumption patterns broken down according to income groups. Such a disaggregation would give us important clues as to which differential food price policies would be most likely to be able, in the face of higher prices for the main staple food, to maintain and improve the calorie-protein intake of the poorest 20 per cent in the rural and urban sectors. The policy choices that would have to be considered require a data-base which, in many countries, is still quite inadequate, and which can only be developed over time.

In short, a great deal more research and analysis will have to be done before it becomes possible to postulate with any degree of confidence what policies and programmes the basic needs model calls for at the local as well at the national level. This brings home quite forcefully that all programmes

---

* Superscript numbers refer to Notes at ends of chapters.

should have a built-in research component that would make possible continuous monitoring and evaluation, and which would ensure flexibility in implementation, and continuous correction and innovation.[4]

Food, Nutrition and Employment[5]

Although the increases in food production achieved by several countries have been spectacular, mainly thanks to the new technologies of the Green Revolution, in many developing countries food production growth rates seem to be beginning to taper off, partly as a result of the slack in existing systems being taken up, but also as a consequence of institutional constraints that are increasingly showing themselves to stand in the way of further rapid increases. The expansion of tertiary and quaternary irrigation networks is an example of necessary development that is often held up by existing land-ownership patterns, small plots and invidious land-tenure practices. The full utilization of potential water resources is sometimes checked by problems of water control. It is becoming obvious that unless social relations in the rural area are democratised, through countervailing policies and legislation, but especially through socially effective organization of the small farmers, the traditional hierarchical structures will continue to exercise important constraints on initiative and productivity.

Land reform and improved land tenure practices should aim at the establishment of higher yielding farm systems through group and co-operative organization, and, where necessary, through consolidation of fragmented land holdings.[6] Such improved farming systems should develop a broader range of food crops, supported by policies aimed at stabilizing food supply, and price policies for the various food crops which will ensure higher production without reducing, but rather improving, the nutritional value of food consumption by the rural landless and urban poor.

Although raising rural productivity is the key to meeting basic needs, increased mechanization has already thrown a number of the landless poor out of work, especially women. Likewise, while price increases in staple foods may provide a stimulus to production, affecting the incomes of the larger farmers especially, its effect on the income of the small farmers and landless labour may be less positive.[7] Even when increased incomes derived from higher food prices do enable the rural poor to pay for their food, the larger farmers might at the same time decide to go for more labour-saving technology, in order to maximize their profits. Policies of this kind, aimed at the small farmer, and where possible at landless labourers, should also encompass support and incentives for intensive home-garden cultivation,[8] small animal husbandry, trees for firewood, and fishponds, which could help in providing additional income, improve the nutritional value of the diet, provide traditional medicinal herbs and restore the ecological cycle at the household and village level.

Labour-intensive public work projects and the stimulation of rural, non-agricultural small businesses might provide the additional income which could compensate for the losses which might be suffered by the poorest groups and the unemployed. In most developing countries there is room for a considerable expansion of such public work projects.

In many developing countries, the combination of poverty and overpopulation has already caused local disturbances in the ecological balance. In these areas there is no prospect of absorbing the doubling of population which has to be expected in the next 25–30 years unless erosion and large-scale damage to the natural environment is halted, and policies to rehabilitate and improve the carrying capacity of the land are carried out. The magnitude of the effort required, as well as its great urgency, provides an opportunity to increase the income of landless labour and the smaller farmers, providing their wages are competitive with those in agriculture.[9] Special efforts should also be made to improve supporting services, like extension and supervised credit facilities,[10] to reach these two target groups, the small farmer and landless labour.

Efforts to increase rural productivity, of course, require an integrated approach to rural development. Effective inter-departmental co-ordination at the village level has proved to be difficult to achieve, except in limited pilot project areas. This raises the question whether effective decentralization and a devolution of power to the lowest possible level of the administrative hierarchy might not ultimately be an essential condition for the effective implementation of the basic needs model. The development of organizational and managerial capability at these levels is at best likely to be a slow and uneven process. To begin with, both administration and grass root participation will have to be experimental in nature and will have to be continuously monitored and evaluated. But it is most important that such policy objectives should be adopted at national level as part of the overall development strategy.[11]

Health

Even in countries which have committed themselves to expanding health care, it has become obvious that it will be impossible to provide conventional health services that would reach the poorest 20 per cent. This is true even in countries where a significant reallocation of the total health funds in favour of rural health has taken place. The effective area covered by health clinics may be no more than between 5 or 10 kilometres in radius, that is the distance a patient can cover on foot. Cost and the shortage of trained medical, technical and other para-medical personnel are the major constraints.

In contrast, the basic needs model emphasizes environmental health and community health care. It is assumed that the provision of safe water and sanitation would significantly reduce the impact of water-borne diseases,[12] and that, coupled with immunization, this would improve the general state of health of the poorest, and the life expectancy, especially of children. Community health care emphasizes the role of community leaders and especially of traditional midwives and traditional doctors using traditional medicine. It also gives a high priority to the training of para-medical personnel with the help of basic training packages at various levels. Primary health promotion and care might even be provided by village workers in the family planning field and in agricultural extension and education. At the same time, it will be necessary to re-orient medical education to the needs of the poorer 40 per cent, without preventing medical schools from keeping up with recent advances in medical science. On top of this, there is the need to inculcate a medical ethos among the young doctors to work and live among, and for, the urban and rural poor. Obligatory national service in environmental and community health care, initiated by various countries, is an important first step. It does not replace the need for an integrated reform of medical education.

Housing

The emphasis on self-help in building and improving housing will require credit facilities on the ground, for building the whole house, or parts of the house (core-house or roof), the development of building associations or co-operatives, the stimulation of local construction and building materials manufacturing firms, using local materials, and site and services preparation in urban sectors. In many cases the villages cannot make use of the credits available, because of uncertainty of land title, incapacity to provide collateral, or legal incapacity to enter into mortgage commitments.

Education

A basic needs approach to education emphasizes functional literacy for all, which (insofar as formal education is incapable of supplying it) will have to be achieved through various forms of non-formal education. Here, the minimum learning package developed by UNICEF[13] might be of great significance, as would be efforts to integrate the literacy effort into other programmes aimed at meeting basic needs. This calls for teaching of functional literacy at the working place, in the family, and the community, with or without the help of teachers.

The next priority will have to be on basic education and the attainment of universal education if not for the whole duration of primary school then for at least three years. Basic education to meet basic needs will require considerable educational reform, covering both the curriculum and teacher training. In addition, the basic needs effort in the field of education will require a re-orientation of higher and secondary education to the problems of the rural areas, and to develop the skills necessary to support this effort.

The basic needs model requires the right development framework at a national level. This should include the adoption of employment and equity as development goals of equal importance to growth. It should also include the determination of multiple growth goals, a commitment to development from the bottom up, to local self-reliance, community and grass roots organization and participation in planning, decision-making and implementation in areas affecting those communities, and a reallocation of total investment, but especially of national funds for health, education and housing, in favour of the lowest 40 per cent income group. Apart from policies and programmes directed towards the target groups, the policy framework calls for:

Land reform (including where necessary the provision of land for home garden cultivation for each household), reform of hierarchical relationships and structures in the rural area, control of land use, and securing for the community, either wholly or in part, the unearned increment created by changes in the value of land.

Abandonment of **price policies for foodstuffs** which traditionally tend to favour the urban population, and the adoption of price policies for different foodstuffs favouring the rural areas in a manner which stimulates food production and increases rural income, while at the same time ensuring improved calorie and protein intake among the urban poor and rural landless labour.

Changing the **terms of trade** between the urban and the rural sector (which traditionally has worked to the benefit of the cities) to the advantage of the rural sector, through alignment of import—and export—duties, and a review of the exchange rate, thus also changing the relative valuation of labour and capital in favour of labour.

Progressive taxation on a broader basis, and reduction of differences in **consumption pattern** between the modern and the rural sector.

Industrial policies giving priority to labour-intensive industries and labour-intensive production processes that are compatible with the requirement of efficiency.

Increase of the number of non-exploitative **linkages** between the modern and the rural sector, once rural institutions and capabilities have become strong enough. Development of a network of agricultural support services, road systems and transportation facilities.

Communications Policy

The effort to meet basic needs obviously requires an unprecedented flow of information into the village, capable of reaching the poorest villagers. It is unlikely that the amount of information a farmer will need could effectively be transmitted through the traditional channels of communication: the village headman, the extension services and the educational system. What is called for is the transformation of the village from a traditional society to an information community, capable of acting and responding creatively to relevant information reaching it, and capable also of reaching out for that information.

Apart from village-level wall-newspapers or newssheets for the tenants, the communications breakthrough that is needed may well require special programmes on TV and radio, and cassettes directed at the villagers. The location of the public village TV set (in countries where this is available), should be such that the programmes are fully accessible during the whole broadcasting period to all villagers including the poorest. This also holds for the village public telephone as well.

Even under the best of circumstances, the communication problem in rural development is a formidable one. There is quite often a wide gap between the needs and aspirations felt at the village level and the assessment of the same needs by administrators and planners at higher administrative levels. The basic needs approach requires new or improved mechanisms for dialogue and interaction—in short, for mutal education—and above all, a greater willingness by planners and administrators to listen respectfully.

Cultural Policies

The efforts to meet basic needs may not have much impact on slowing down the movement towards the cities, unless cultural life within the village can be improved. If TV is not to be an instrument which increases the attraction of the cities, it will have to be domesticated to serve the cultural needs of the village. Decentralization of programming would open the way for locally-produced programmes which would stimulate local creativity, and would revive the community's capacity for self-enjoyment. It would also halt the long process of cultural impoverishment which has afflicted village life in many developing societies. Decentralized programming, linked, for instance, to village competitions in music, theatre, dance, literature, sports, children's games and folklore, painting, sculpture, wood-carving and other crafts, would be an important means of combating the passivity that usually characterizes TV audiences.

At a different level, it is impossible to overestimate the difficulties of helping the villager to break out of the traditional hierarchical social structures which have for so long imprisoned him. Changes in the local distribution of power, through democratization, are of course, one important precondition. But many traditional attitudes, among rich and poor alike, are rooted in transcendental preconceptions of the social order. Individualization, the growth of personal self-reliance and initiative, and the realization that to try to improve one's material conditions and those of one's family are important elements on the road to emancipation. And cultural policies aimed at stimulating those qualities and values have great developmental significance.

Research and Technology Policies

As stated before, all basic needs programmes will have to have a built-in research component in order to make possible continuous monitoring and evaluation, adjustment and innovation. In many developing countries, however, the growth of research capability has not matched the expansion in the use of new technologies. In addition, a national research policy supporting the basic needs approach will have to include research and development aimed at bringing indigenous technology, including agricultural implements, up the next step of the technological ladder, without taking the improved product out of the reach of the small farmers. Developing countries will also have to direct a larger part of their research capabilities to those problems that are especially important to the poorest part of the population, or which at least are neutral in their social impact.

Many of the basic services use at present a technology that is still too expensive to reach the poorest people. New, cheaper technologies will have to be developed that are replicable on a large scale. In addition, special attention should be given to the development of one other type of research and analytical capability. It is extremely important for a country committed to the basic needs approach to be able to make periodic assessments, not of where it intended to go, but of how far it has come and of whether it should pursue or adjust the policies it is pursuing.

Energy Policies

The basic needs model raises particular questions about the supply of energy to the poor.[14] Rural development will require increased energy use for irrigation, fertilizers, draft power for cultivating the land, as well as small machines for agricultural processing, and small scale manufacturing; and also, of course for cooking and heating.

While in some areas rural electrification may be the answer, in isolated areas, or those that have a widely dispersed population, the development of alternative energy resources is an absolute necessity (water, wind, use of waste materials, bichemical energy conversion through utilisation and upgrading of the ecological cycle within the village, and photovoltaic devices.[15] For some time firewood is bound to remain the major energy resource of the poor. This will require the development of programmes designed to make better use of what is available; both by reducing waste as well as increasing supply (reforestration and establishing village wood lots). In all these areas, links will have to be made with research and development abroad, especially with those concerned with the soft technologies in energy.

Administration Policies

Before the advent of the basic needs concept, state bureaucracy was generally considered to be the main instrument for implementing the development plan, and for mobilizing and guiding the population in the development effort. The basic needs model, in contrast, with its emphasis on development from the bottom up, puts a premium on the organizational and management capacity of rural communities, as well as on the development of co-operatives and other forms of organization. It means the adjustment of traditional hierarchical patterns and patron-client relationships to more modern, more democratic forms of social organization, capable of tackling new problems. All this runs directly counter to the conventional bureaucratic approach to the village, which tended to strengthen those traditional structures. It means, in effect, a quantum jump from paternalism to emancipation, requiring fundamental changes in attitudes on the part of the administrators as well as of the people at large.

The transition will not be an easy one, particularly as any central bureaucracy tends to overlook the inherent limits to its role as the instrument of development. For example, at some point in the development process the initiative and responsibility must shift to the farmer and the rural community if further progress is to be made. Secondly, bureaucracy is incapable of policing itself effectively in the absence of countervailing social forces. Thirdly, despite all good intentions from the top, particular policies will inevitably be twisted because of pre-existing cultural and administrative patterns, as well as because of the interplay of different interests.

Another reason for difficulty in bringing about this transition is the unavoidable loss in efficiency resulting from the handling of particular developmental tasks by inexperienced organizations and institutions. While it is desirable that these young institutions be given the time to make their own mistakes, central govenment is rarely willing to provide it. But it is only in freedom that these institutions can learn, developing the skills and the self-confidence and self-discipline essential to their further development. It is therefore likely that only a firmly committed government, capable of maintaining continuous pressure on its bureaucracy, will be able to bring about such a transition. The basic needs model also requires the administrative system to adjust to the role non-governmental organizations and institutions can play at the grass root level. It will have to learn how to nurse them without suffocating them in its bureaucratic embrace.

Changes are also needed in the training and education in public administration. Today's attitude towards public administration as a sanitized, insulated, technocratic enterprise should be replaced by an understanding of how public administration is part of a whole network of social interactions. Just as important as effectiveness and efficiency in public administration is an understanding of the pitfalls that face an administrator in the implementation of policies. We will need administrators who have a sense for the point at which efficiency must give way to justice, paternalistic guidance must give way to self-initiated experimentation, and who know when not to interfere.

Legal Policies

Land reform constitutes in itself, of course, a major piece of legislative work. The land reform law will have to cover adequate monitoring mechanisms and procedures for protection of the small farmers as well. The establishment of co-operatives and other rural associations, of rural credit institutions, of the housing programme and of rural businesses, requires a legal infra-structure in the rural area which in many developing countries does not exist.

The basic needs model also calls for a systematic extension of the judiciary to the village level. A revived rural sector needs a greater capacity for resolving conflicts. Traditionally, it has been the village headman who adjudicated conflicts and acted more or less as a justice of the peace. In many countries he now has become involved in implementing development programmes, and, as such, a party to many disputes. Traditional judicial mechanisms have, therefore, broken down, creating the need for a new means by which the villager can seek legal redress for injustice.

Political Dynamics and Ideological Implications

This incomplete list of national policies which constitute the development framework required for the basic needs model serves to bring out that the model is not yet a substitute for a development strategy. It expresses a particular emphasis, and a particular approach to the development process. But it becomes meaningful only when it is set within a framework of regional and national development policies that are capable of coming to grips with such a change in emphasis.

Simply listing these national policies also makes it clear that fundamental changes are needed. Changing the balance between the urban and the rural sector, between the centre and the periphery, reducing consumption among the élite as part of the reallocation of total resources in favour of the countryside, opening the door of political life to the hitherto disenfranchised or socially ineffective sectors of the population, and the broadening of the political power-base that this entails, together amount to a structural transformation and a fundamental change in the distribution of economic and political power.

Bringing about such changes is a big political risk. But the risks involved in *not* dealing with the problems of employment and equity may in the long run be even greater. There is, therefore, trade-off involved between risk today and in the future. Be that as it may, it should be realized that despite the often authoritarian character of governments in developing countries, the political systems remain fragile. (Maybe one should put it the other way round: they are authoritarian because their power is so fragile.) In several countries, the political system has proved to be incapable of making the adjustment from growth to equity goals. There are evident limits to the capacity of a political system to make a fundamental adjustment of this kind within a short time.

The experience of a number of developing countries in South America has shown that, stretched beyond a certain point, democratic political systems can collapse through the breakdown of national consensus and social cohesion, plunging the nation into a tailspin of increasingly irrational violence, from which it seems almost impossible to recover. Such a collapse can take place both when the problems of equity continue to be disregarded, but also when the effort to address the problem has been made, but has failed. The basic needs model therefore represents not merely a shift of emphasis in the development effort; it is a profoundly political enterprise often evoking primordial fears and violent reactions. It means a slower development of the modern sector of the economy with its own legitimate interests, and the postponement of the hopes of labour to improve its position. It is also clear that the full realization of all economic, social and political implications of the basic needs model invites a degree of state intervention and control which only a totalitarian state would be capable of generating and maintaining. This would undoubtedly destroy the opportunities of development for the poorest part of the population together with the growth of their organizational, managerial and self-assertive powers through which a greater range of choices will open up to them. These would be crushed under the weight of either a technocratic or revolutionary bureaucracy, be it from the left or from the right.

On the other hand, without such a set of national policies, the basic needs model would soon be reduced to a scheme for the supply of food and basic services, with little change in the condition of powerlessness and dependency of the poorest. In other words, freedom is in itself a basic need, if the meeting of basic material needs is to lead to emancipation and self-reliance.

We are faced here with a crucial contradiction inherent in the development process. On the one hand is the need for strong central power and economic direction from the top. On the other hand, the building of a participatory society requires freedom. Local autonomy, self-reliance and socially effective participation at the village level are inalienable parts of that freedom. No developing country, including China and Tanzania, Brazil or Kenya, seems to have been capable so far of conciliating these conflicting demands satisfactorily. It would therefore seem that the importance of the basic needs model does not lie so much in itself, but in it being a potentially important building block in a democratic theory of development, still to be formulated.[16]

The international debate about the basic needs model so far has not satisfactorily clarified how it would fit into societies based on different economic systems, a socialist system, a classical market economy or a mixed economy. While the model is usually taken as an ascent to freedom, its programmatic features lend themselves quite easily to paternalistic or authoritarian methods of implementation. It is not impossible to imagine nations explaining their transformation into totalitarian societies in terms of their commitment to the basic needs model. The model itself remains silent about these ideological questions, and provides no clue as to how the totalitarian pitfall could be avoided.

Uncertainty about the full implications of the basic needs model, and about how committed the industrial countries are to it pose external dilemmas for governments of developing countries committed to the approach. The commitments made by the OECD countries at the World Food Conference to establish a fully adequate grain reserve, to provide 10 million tons of grains for food aid conventions in the present UNCTAD negotiations, to pursue nutrition programmes, to expand development assistance of food, and to resolve current food trade problems, have so far not been honoured. Since food production is a central element in the basic needs model, this failure has created doubt about the seriousness of the commitments of these countries to the basic needs model.[17] The questions that then arise in developing countries include the following:

It has become obvious that the basic needs approach to development requires structural changes both in the developing countries and in the industrial countries as well. Without such changes the basic needs approach stands little chance of success. To what extent then are donor countries really prepared to accept the transnational economic implications of the basic needs model, including stable commodity prices, access to the markets of the developed world for goods produced by developing countries, and worldwide economics on depletable resources?

The basic needs model calls for industrial policies which at least to begin with insulate small rural enterprises using local materials from the greater competitive power of more efficient modern sector industries, including those in the developed world. Policies aimed at redressing the imbalance between rural and urban sectors would also affect private foreign investment. Even if such policies were to affect only new investments, would the developing country concerned lose its attractiveness to private foreign capital? In light of such policies, would donor countries not lose interest in maintaining adequate aid levels? And would foreign experts lose interest in working in developing countries where an affluent lifestyle is considered unacceptable?

In its initial stages the basic needs model proper requires mainly domestic funding. It must use cheap, locally available materials, develop rural institutions and increase rural organizational capability at the village level. Its foreign exchange requirements at that stage are therefore quite small. It is only in its later stages that the basic needs model would require a much larger financial input, and could even become itself a foreign exchange earner. Nevertheless, we see more and more donor countries jumping on the basic needs bandwagon, without much questioning as to how much foreign aid the model can absorb. In a number of countries this is already causing long delays in the aid pipeline.

Do donor nations think that through better management or greater efficiency which they could insist on or themselves provide, that capacity could be increased within a short time? How do donor nations visualize their role in the effort to increase rural absorbtive capacity and rural productivity? Is it possible for foreign bureaucracies to push for the structural changes at the local level which this requires? Here, more than in working at the national level, is where interests clash, where tremendous communication problems loom ahead, and where cultural differences are bound to be the sharpest, and a constant source of conflict. Are the donor nations prepared to accept the political consequences of such a deep intrusion by them into the life of another people? History shows that only through occupation after military conquest, or through the colonial relationship, could a foreign power hope to bring about social changes that suit their own perceptions and values. It is impossible to think that it is this what the donors have in mind. And finally, would any developing country be prepared to accept foreign aid if it is combined with this kind of a foreign presence in their rural areas?

It is obvious that the basic needs model brings the donor face to face with the most intractable of all problems of underdevelopment, which will need much time and effort to overcome. Are they willing to commit themselves to such a long and sustained effort? Will their national constituencies, when social development turns out to be much more difficult and slow in producing tangible results, not lose interest even more quickly than was the case with the earlier concepts of aid? Can the developing countries count on such a commitment on the part of the donor countries for at least 20 years? Such a long term commitment would only be feasible if it were linked with real and important interests of the industrial countries. Still the question remains.

The basic needs model requires a much more sophisticated understanding of the development process, but very little effort is being made in the donor countries to generate such an understanding. On the contrary, there seems to be an increasing reliance on the humanitarian argument. But a humanitarian emphasis reduces the basic needs model to a relief operation, disregarding real development needs, and this would leave the poor even more permanently in a situation of powerlessness and dependency.

The earlier discussion shows that there are a number of development requirements in support of the basic needs model: area development, roads, irrigation, transportation facilities and so on. Would donor countries be interested in meeting these needs? Would they be interested in making available the processing plants, the cold storage facilities, the motorized fishing boats and the trucks, if these are to be owned as soon as possible by farmers' and fishermens' co-operatives? This is relevant, because some spokesmen in the industrial countries seem to believe that the basic needs model only requires small mini-projects.

Speaking more broadly, would the donor nations be interested in committing aid funds to the modern sector developing countries, which might otherwise suffer because of the emphasis on basic needs? Or is the unspoken intention to leave modern sector development entirely to private foreign investment and to the government? In other words, is it possible for the donor nations to envisage providing capital goods, modern technology and science for the modern sector, instead of concentrating on rural development with its inherently limited absorptive capacity? Such a basic reorientation of the role of foreign aid would constitute convincing testimony to the seriousness of the donor countries in supporting the basic needs approach.

Such a concept of foreign aid would go a long way towards overcoming the fear of many developing countries that the emphasis on basic needs is only a manifestation of a desire on the part of the major industrial countries to maintain the developing nations as better fed, better housed and better schooled, but still as second class pastoral societies. It would also counteract the other fear, that the basic needs emphasis is simply a way of reducing aid levels while feeling virtuous about it.

Technology transfer, adaptation and development, in conjunction with social organization and development is very much culture-specific. It requires an indigenous capacity to develop new knowledge, new skills, and new technological capability relevant to development needs. While it was possible and rational for donor countries and international agencies to have teams of experts moving about from one country to another, capable of building infrastructure in any country, the knowledge and skills required in the basic needs context will have to be developed on the ground by national research institutions and country specialists. To what extent would donor countries be interested in supporting such efforts through special training programmes, financial aid and technical assistance?

A mutual commitment of both donor nations and developing countries to the basic needs model would inevitably lead to a major reorientation of the science and technology production system in the industrial world. So far they have largely remained insensitive to the needs and the dilemmas of the Third World. It will be necessary for industrial countries to look harder for high technology which could be utilized by the small farmer as well as the richer one, and which could be adapted or copied cheaply in the developing countries. Further development of photovoltaic cells could be one such example. A recent report by the US National Academy of Sciences[18] identifies among areas of high priority the efforts towards making plants more responsive to nutrients in the soil and more resistant to pests and adverse climatic conditions.

Social development and the building of rural institutions is slow. How will this affect the operations of the bilateral and multilateral donor institutions? It might imply, for example, the need for more powers of decision for the aid representatives of donor countries and international agencies in the developing host country. It may also mean adding social scientists to their staff at both the field level and in the decision-making process at headquarters. Are donor countries and agencies willing to make these changes? Are they willing to allow a more autonomous role to domestic institutions within the host country? Are these agencies prepared to adjust their monitoring, evaluation and administrative procedures to enable them to work with the non-governmental organizations at village and community level? Would they be willing, for instance, in projects of this kind, to go as far as to accept forms of joint post-audit control?

To what extent are donor nations prepared to accept the full ideological implications of the basic needs model? Is it assumed that the model can be financed solely from growth, and that therefore no redistribution of productive wealth, except land-reform, would be required? But what if the necessary 6 8 per cent were to be unattainable? And what if more drastic redistribution policies, like nationalization, were to be resorted to?

Would donors still be prepared to fill the remaining resource gap? Or would they respond ideologically, and cut aid? The historical record of some major donors makes it risky, to say the least, to discount the possibility that today's friendly donor might turn into tomorrow's enemy.

This paper has tried to show how many of the domestic and international dimensions of the basic needs model have so far remained unexplored, and how urgent a fuller exploration is. Such a clarification can only be achieved through a serious and continuing dialogue between the development communities of the industrial and the developing nations, sustained over a long period of time.

Even if it is assumed that the will is there, the obstacles are formidable. On the part of the industrial countries, these may stem in part from their preoccupation with their own problems and opportunities, but also from a basic insensitivity to Third World problems and dilemmas. On the other hand, we should realize that in rapidly changing transitional societies, the development effort is just one among many processes of social change, although it may be the only one over which the government has some degree of control. All these processes interact; they fuse and collide; they change and modify each other in the process, moving in constantly changing configurations. We are beginning to realize how little we really know about development, but the magnitude, violence and inhumanity of some the convulsions through which some developing countries have gone also shows how limited is our comprehension of these historical processes, and how even more limited the capacity of governments to control them. However strong they are, however committed to development, governments soon find out how much more complex are the problems that they have to face and how well-nigh impossible it is to capture historical processes in neat models and flow-charts. For the dialogue, this creates difficult problems of explanation and credibility.

After all, how does one explain across the deep divide of culture things like the trade-offs between effectiveness and freedom, efficiency and justice, objective urgency and the internal rhythm of social growth? And, having had to thread one's way through a barely reconnoitred minefield, how sure can one be that after so many detours and compromises one has not lost one's way? The possibilities on both sides for misunderstanding, disillusionment, misjudgement and condemnation, are therefore numerous. In addition, the number of people available in the Third World with the necessary time and competence to carry out the dialogue is extremely limited.

In the end, the massive intellectual effort and the staying power which a continuing dialogue requires can only be generated and maintained when three other factors are present. First, mutual trust on a personal level between the participants in these discussions. Secondly, mutual faith in the basic qualities of the other people, despite the inevitable setbacks and disappointments. And thirdly, faithfulness; a willingness to suspend final judgement for a long period of time once a common commitment has been made. These are rare commodities in today's world, but it may not be impossible to find them. The fate of the poor on this globe, as well as our common survival, may well depend on it.

Notes

1. *"Catastrophe or New Society?" A Latin America World Model* by Amilcar O. Herrera *et al.* International Development Research Center, Ottawa, Canada, 1976.
   Employment, Growth and Basic Needs: a One-world Problem. International Labour Office, Geneva, 1976.
   The Changing World Order and the World's Poorest Billion: A Fresh Approach to Meeting Essential Human Needs by James P. Grant, President, Overseas Development Council, Washington DC. A paper presented at the 25th Pugwash Conference in Madras, India, 12–19 January 1976.
   "The Planetary Bargain." Proposals for a New International Economic Order to Meet Human Needs. Report of an International Workshop convened in Aspen, Colorado, 2 July–1 August 1975.
2. *Meeting Basic Human Needs: within Harmonious Environmental and Development Objectives.* A Feasibility Study, by John McHale and Magda Cordell McHale. Houston, Texas, February 1977. Also by most UN Specialized Agencies.
3. *Redistribution with Growth* by Hollis Chenery, Montek S. Ahluwalie, C. L. G. Bell, John H. Duloy and Richard Jolly. International Bank for Reconstruction and Development, USA, 1974, contains an interesting discussion of the political framework. The topic, however, still awaits a full analysis.
   See also: *Human Development and Social Change* by Marshall Wolfe. ECLA, Social Development Division, E/CEPAL/98, October 1975.
4. *Ethics and Programmatic Thinking about Rural Welfare* by Kalman H. Silvert. October 1972.
5. For this section the author has drawn heavily on the excellent study, *Meeting Basic Human Needs: within Harmonious Environmental and Developmental Objectives.* A Feasibility Study for the UN Environment Program, by John McHale and Magda Cordell McHale. University of Houston, Texas, February 1977, pp. 118–158.
6. *Food Problems, Unemployment, and the Green Revolution in Rural Java* by William L. Collier. Prisma, January 1978.
7. William L. Collier, op.cit.
   Also: *Economic Development and Shared Poverty among Javanese Sea Fishermen* by William L. Collier, Harjadi Hadikoesworo and Marie Malingreau. Prepared for the Agricultural Economics Society of South East Asia's biennial meeting, Philippines, 2–6 November 1977.
   The Changing Structure of Agricultural and Non-agricultural Employment Opportunities in Rural Java: Future Prospects and Policy Considerations, with Special Reference to Handicrafts and Home Industries by Benjamin White. Paper prepared for the Seminar on Rural Industrialization, Lembaga Pengembangan Masyarakat, Gajah Mada University, Yogyakarta, 22–24 December 1977.
8. *The Javanese Home-gardens as an Integrated Agro-ecosystem* by Otto Soemarwoto. In "Science for Better Environment", Proc. International Congress Human Environment (HECS), Science Council of Japan, Tokyo, 1976, pp. 193–197.
   New Approaches in Community Nutrition Programs by Dr. Sajogyo. Paper at SEADAG Seminar on Food and Nutrition, 28–30 April 1975.
9. *Food Problems, Unemployment, and the Green Revolution in Rural Java* by William L. Collier, Prisma, January 1978.
   Also: *Income, Employment, and Food Systems in Javanese Coastal Villages* by William L. Collier, Harjadi Hadikoesworo and Suwardi Saropie. Ohio, University Center for International Studies, Southeast Asia Program, 1977.
10. *Credit for Small Farmers in Developing Countries* by Gordon Donald. Westview Press, Inc., Bouldar, Colorado, 1976, pp. 261–271.
11. Foreign assistance could play an important role in support of such rural institution building, as well in helping underwrite the inevitably high risk credit insurance connected with the operation of these fledgling institutions.

12. See: *Prospects for Rapid Decline of Mortality Rates in Java* by Terence H. Hull and Jon E. Rohde.
13. The minimum learning package comprises the following six elements: positive attitude towards cooperation; functional literacy and numeracy; a scientific outlook and an elementary understanding of the processes of nature; functional knowledge and skills for raising a family and operating a household; functional knowledge and skills for earning a living; functional knowledge and skills for civic participation.
    See: *New Paths to Learning. For Rural Children and Youth* by Philip H. Coombe, Roy C. Prosser, Manzoor Ahmed and Barbara Baird Israel, Editors. Prepared for UNICEF by International Council for Educational Development, USA, October 1973, pp. 14–15.
14. John McHale, op.cit., pp. 159–182.
15. *Energy for Rural Development*. National Academy of Sciences, 1976, p. 12.
    See also: *Rays of Hope. The Transition to a Post-petroleum World* by Denis Hayes. Worldwatch Institute, USA, 1977, pp. 155 ff.
16. *In Search of Theoretical Room for Freedom. North American Social Science Thinking about Latin American Development* by Kalman H. Silvert. Paper for presentation at Sixth National Meeting of Latin American Studies Association, Atlanta, Georgia, 25–28 March 1976.
17. Statement of the *Amsterdam Symposium on Food and Basic Needs,* organized by the Netherlands Committee for a New International Order and the Centre for World Food Studies. Amsterdam, 26 February 1978.
18. *World Food and Nutrition Study. The potential contributions of Research.* National Academy of Sciences, Washington DC, 1977.
    See also: *Rural Development in the Third World and the Impact of Fertilizer Demand* by James P. Grant. Overseas Development Council, Washington DC.

*Closing the Widening Gap**

B. M. UDGAONKAR†

UNESCO's initiative in launching its programme on research and human needs has been valuable and timely. Its purpose has been twofold: first, to contribute to the acceleration of the process by which the world's scientific community and research results are brought to bear on the resolution of the most urgent problems facing mankind; and secondly, to identify trends in the natural and social sciences which have a bearing on meeting basic human needs for all people in the world.

The first of these questions, which has to do with the motivation or value system of the scientific community on the one hand, and the political will as represented by national or international decisions to attack these urgent problems on a war footing, on the other, is to my mind the most crucial question. Once the scientific community realizes that these problems are worth urgent attention and/or once the political decision is taken to mobilize scientists and resources for action on an emergency footing, it should not be very difficult to deal with the second question. In fact, the priority areas have been listed by several UN bodies. And there are several examples in recent history to show how scientists can be successfully mobilized for fulfilment of politically defined goals—e.g. war-time atomic and radar programmes and more recently, the Man-on-the-Moon programme.

My remarks will therefore largely be confined to the first of these questions. This question again may be approached either from a global point of view, or from a national or regional point of view. While the prime focus of my discussion is on national efforts, it is hoped that it will also suggest possible regional or international action.

It has been increasingly realized during the last few decades that progress in science and technology has for the first time in history made available to man the tools by means of which poverty, ignorance and destitution could be banished from the surface of the earth. On the other hand, one is faced with the unfortunate situation wherein the number of people below the poverty line, and the number of illiterates (the two groups have a very considerable

*First published in *New Quest,* Vol. 1, No. 1, May 1977.

†Professor B. M. Udgaonkar is at the Tata Institute of Fundamental Research, Bombay, India.

overlap with each other) show no signs of decrease over vast areas of the globe, and the gap between the poorer and the richer countries is increasing every year—and this is happening in spite of the fact that the expenditure on research has grown exponentially during the post-war period.

It may be worthwhile to emphasize that the basic or minimum human needs of food, shelter, clothing, education, health and work represent an area of urgent social action, not merely on compassionate or humanitarian grounds, or because their neglect in the so-called Third World is likely to create an explosive situation for the richer part of the world, but also because their satisfaction is a minimum prerequisite for the restoration of human dignity to a large majority of the world's population, and for giving a real meaning to democracy.

While research and action—in fact, action-oriented research—on the problem of meeting the basic human needs is a moral responsibility of both the richer and the poorer part of the world, I shall address myself largely to the role of scientists (including social scientists) and their institutions in the Third World (India in particular) in relation to the fulfilment of basic human needs. I believe that the widening gap between the so-called developed and developing countries, in spite of several "aid" and "technical assistance" programmes, underlines the importance of self-reliance by the countries of the Third World.[1] Self-reliance is itself a basic need for any nation or region which wishes to make a serious attack on the problem of meeting the basic human needs of its population.

Modern scientific research, especially organized research, has a history of only a few decades in the poorer part of the world. In India, for example, while outstanding contributions to basic research were made by men like J. C. Bose, C. V. Raman and S. N. Bose in the pre-independence period in spite of heavy odds, facilities for research were very meagre until the mid-fifties, and it is only during the last two decades that several scientific institutions of high standing have emerged. The growth of these institutes and research centres, whether inside or outside the universities, represents the creation of a scientific capability of high order, and this scientific capability can be and must be brought to bear upon the pressing problems of development—or of meeting the basic human needs in the first instance. One may naturally ask why the potentialities represented by such a scientific capability, and the technological capability which has also grown side by side, are not yet adequately utilized for a frontal attack on the basic problems. This may be asked within a country, as also on a world-wide scale.

Part of the reason for the mismatch represented by the slowness of action on this problem is, of course, political. One may even ask if the relatively small élite which often holds the reigns of power in the poorer countries is

---

* Superscript numbers refer to Notes at ends of chapters.

interested in foregoing some of its privileges in order to make scarce resources available for a massive attack on the problems of poverty and ignorance. This no doubt calls for a political decision. But political decisions (especially in an open society) are not made in a vacuum, and one may ask if the scientists (or intellectuals in general) are sufficiently concerned about these problems and ready to play their role if called upon to do so.

In this connection, one comes face to face with a certain dichotomy of values among intellectuals. When discussing basic human needs, they will agree that the basic human needs are food, shelter, clothing, health, education (at least "functional literacy") and work, or employment, and that it is a pressing problem to make these available to the people, 60–70 per cent of whom may be below the poverty line in a country like India. On the other hand, when it comes to basic research, it is defined as research at the frontiers of our knowledge, aimed at an understanding in depth of phenomena—whether physical, biological or social. The prestigious research is basic research, as remote as possible from any applications, not motivated by any possible use, and not related to any action, especially social action, or steps directly aimed at providing the basic needs to the underprivileged sections of the society. Somehow these problems are not considered sufficiently challenging intellectually by the academic community.[2]

There is therefore a considerable gap between prestigious research and basic human needs. In the relatively closed systems such as those which exist in certain countries, such elitism is not tolerated and the gap is sought to be closed by demanding that every academic devote a certain fraction of his time to socially relevant work, including work on farms or in industry. In an open society such as that in India, however, on the one hand, a large fraction of the scientists and technologists—especially the more senior ones who create the ethos—are trained in the richer part of the world, and continue to look to this outside world for the choice of their research areas, and for approbation and recognition (not to mention the attractive benefits of travel and stay in the richer countries; in some cases, a permanent stay contributing to the so-called brain-drain). The scientific community has yet to evolve its own internal standards of "relevance" and of recognition of work of merit, including work related to the solution of urgent national problems, in particular, problems relating to the satisfaction of the basic needs of the vast majority of the population. On the other hand, the governments has also been slow in recognizing the strength represented by the existence of a substantial scientific and technological infrastructure, to the creation of which it has itself contributed, and the role this infrastructure could play in the Basic or Minimum Needs programme if the political decision was made to utilize it.

However, the scientific academic scene presents a growing number of examples where institutions and individuals have engaged themselves in nationally important programmes, programmes of developing self-reliance in critical fields or programmes directly related to the other basic needs. I shall now briefly describe a few examples of such efforts to remove or avoid the above mismatch, with the hope that they may suggest some models for action.

While these examples are all drawn from the Indian scene, I believe they may have some relevance elsewhere.

As a first example, I would like to consider the radiating influence that may be exercised by a centre of fundamental research in basic sciences. The Tata Institute of Fundamental Research (TIFR) was founded in 1945, with the avowed objective of providing conditions where first-rate research work in frontier areas may be carried out within India. This continues to be the primary objective of the institute and the institute is known for its contributions to basic sciences in several areas. In addition, however, over the period of three decades of its existence, through the expertise generated by its research programmes and a certain sensitivity to national problems which it has always encouraged, it has helped the growth of a variety of developmental programmes—in applied sciences related to national objectives and in education. The institute served as the cradle of the Indian atomic energy programme and not only provided a large number of initial staff and the early leadership for it, but the control system for India's first nuclear reactor was also built in some old war-time hutments on the site of the institute. The instrumentation group set up at the institute in the early years, to design and develop electronic instruments, became in course of time the Electronics Division at Trombay, and still later developed into the Electronics Corporation of India, a premier public sector undertaking which now manufactures a wide variety of electronic instruments, including computers, and takes legitimate pride in its success based on total self-reliance. One may also mention that the Bhabha Committee on Electronics laid the foundation for the development of electronics in India, and when the Electronics Commission of the Government of India was established in 1971, this Commission again drew upon personnel (including its Chairman) from the TIFR for its work and for carrying out some of its major projects.

Early work on the development of accelerators like the electron linear accelerator created an expertise in microwave devices and instrumentation, which is finding national applications. The radio-telescope is a tool of basic research; but once it was designed, fabricated and commissioned indigenously, it not only gave rise to some exciting astronomical results, but it also created the base for further work relating to steerable antennas needed for communication systems and radars in the country. Research programmes in the field of cosmic rays have led to applications in geochronology, archaeology, hydrology and meterology, and resulted in the creation of a national facility for C-14 dating. Research in computer sciences has led to the creation of a National Centre for Software Development and Computing Techniques at the institute. The Dental Unit at the institute, which ordinarily would be expected to confine itself to service functions, is engaged in an important programme of oral cancer research.

The institute has also been playing a constructive role in its immediate surroundings in the field of education—from the primary school level to the university level. Involvement in science curriculum development programmes at school level has led to the creation of a Homi Bhabha Centre for Science Education at the institute. This centre gives a special emphasis to programmes suited for the poorer schools in the city and in rural areas.

When the Space Applications programme took shape, a senior member of the institute was drafted to head the Space Applications Centre and the Satellite Instructional Television Experiment—a programme with a focus on the use of TV for rural education. Members of the institute have taken an active part in creating some of the software for the programme. Members of the institute have also been active on national bodies like University Grants Commission, National Committee on Science and Technology, and so on.

All these activities have grown at or from the institute without affecting its primary focus, which continues to be basic research in frontier areas.

The example of TIFR underlines the importance of creating a centre of excellence in basic sciences in any country which wishes to harness science for development. The intellectual atmosphere, the training of the mind, and the generation of expertise which go with such a centre, and the confidence that comes from doing competitive research in frontier areas *under local conditions,* are an invaluable asset for manpower training in general, and leadership training in particular, As Homi Bhabha pointed out at the inauguration of the new buildings of TIFR in 1962, the support of fundamental research, and of an institution where such research can be carried out effectively, is of great importance to society.

> "First of all, and paradoxically, it has an immediate use in that it helps to train and develop, in a manner in which no other mental discipline can, young men of the highest intellectual calibre in the society into people who can think about and analyse problems with a freshness of outlook and originality which is not generally found . . ."[3]

The Third World, as mentioned earlier, represents countries with a wide spectrum of scientific and technological capabilities. While some of these countries have already built a certain infrastructure and capability, or are in the process of doing so, a large majority of the Third World countries are either too small to build a viable infrastructure of their own, or have been so much neglected by their erstwhile colonial masters that they will take considerable time to build such an infrastructure. On the other hand, given the fact that the life-styles and approaches to industrialization and development of these countries may be quite different from those in the "developed" countries, (which are themselves wondering whether they have followed the correct path of "development") the solutions to their problems are not likely to be found through research programmes located in the "developed" countries. It would therefore be desirable in this situation, to adopt a regional rather than a country-wide approach, in creating centres of excellence in carefully chosen areas of basic research. An enlightened international scientific community could help in this process, taking care to respect the cultural values and developmental styles of the region concerned and to ensure that the resultant effect of any international co-operation is to contribute constructively to the development of indigenous capability. The International Centre for Insect Physiology and Ecology in Nairobi is an effort of this kind, and Djerassi[4] has proposed an extension of this model.

The second example I wish to refer to is an institution of a completely different character. The Gandhigram Rural Institute of Higher Education near Madurai in South India is a part of a complex of educational and community service institutions, which have together been involved in teaching, training, production, extension, and research activities, in an area covering 26 villages. While the institute does not yet have the formal status of a university, its diplomas are treated as equivalent to university degrees by the Central and State Governments in India. Unlike the conventional universities, teaching is integrated with extension work and action-oriented research in the academic programmes of this institute at both the undergraduate and the post-graduate level. The Gandhigram experiment in integrated rural development (public health and sanitation, child-care, family-care and family planning, building of roads and houses, etc.), with an educational complex as the nucleus, has now been going on successfully for about 25 years, and shows what an educational institution with a social purpose can do to contribute to developmental problems of the community around it.

The council of Scientific and Industrial Research (CSIR) of the Government of India runs about thirty Research and Development laboratories covering a whole spectrum of scientific disciplines—physics and chemistry, food technology and drugs, petroleum and fuels, mining and metallurgy, roads and buildings research, environmental engineering, electronics and scientific instruments, mechanical engineering, leather research, etc. Its concern for a direct utilization of the results of research for the socio-economic progress of the country led it in the early 1970s to formulate a plan for the all-round development of a backward district in the State of Andhra Pradesh and another in Maharashtra.

> "It is hoped that by pooling the scientific and technological inputs which the CSIR and other agencies will bring in and the financial and

manpower resources available from the State Government, financial institutions and local entrepreneurs, it would be possible to develop various sectors of the economy in this district rapidly through application of science and technology within a given time frame."[5]

The Indian Institute of Science, Bangalore, is one of the premier scientific and educational institutions in the country. "In response to the increasing concern over the strong bias shown by major educational, scientific and technological institutions towards urban problems", it created in 1974 a cell for the Application of Science and Technology to Rural Areas. This cell is intended (a) to catalyse the development/testing of village-oriented technologies on the Institute campus; (b) to establish an Extension Centre amidst a cluster of villages near Bangalore; and (c) to accomplish and monitor the transfer of the developed/tested technologies to rural areas through the Extension Centre and through established rural development agencies.

The first projects initiated included projects on the development and exploitation of locally available energy sources such as windmills and biogas plants; on the mechnical engineering design of bullock-carts, hand-pumps, bicycle-drawn trailers and vehicles; on small scale industries to exploit natural resources and agricultural wastes; on rural housing. In order to enable institute students to participate fully in this programme, ASTRA problems have been offered as M.E. projects, as part of the academic requirement of the institute.

The appropriate technology unit (ATU) at the Indian Institute of Technology, Bombay, represents another effort, prompted by similar motivations. Mention should also be made of the efforts of a group at Indian Institute of Management at Ahmedabad, to serve as catalysts of a development process in a block of 200 villages in the State of Rajasthan.

An interesting programme of student involvement in community development, under the umbrella of a university, was launched by the University of Bombay in 1973. The programme started as a "National Graduate Scheme" (NGS), with a group of about 15–20 graduate volunteers from various disciplines, who were placed in selected villages around Bombay. The aim was, on the one hand, to study the appropriateness of university education for the present social environment and, on the other hand, to understand the practical local problems—in agriculture, finance, community living, etc.—and to try to contribute to their solution.

The experience has been quite encouraging. The students not only realized the weakness and in some cases the irrelevance of the bookish education they had received from the point of view of understanding the social, economic and political institutions in an operational manner; but exposure to realities of rural life—such as the stifling hold of the vested interests over the people and community resources, and the prevailing sense of resignation and helplessness among rural folk—made the volunteers lose their innocence and sensitized them to the need for social action.

My final example refers to a completely different kind of participation by the academics in universities and research institutions in a programme of rural development, with special reference to science education. Kishore Bharati and Friends Rural Centre are two voluntary organizations in the Hoshangabad district of Madhya Pradesh, engaged in an integrated rural development of an effective science teaching programme in sixteen existing middle schools of that area. The teachers in these schools had little or no background in science, and almost all of them lacked a university degree. For the first orientation course for teachers from these schools in 1972, they enlisted the co-operation of members of the All India Science Teachers' Association Physics Study Group and of several scientists from TIFR (where the leader of Kishore Bharati, a molecular biologist, had worked for some years before he founded Kishore Bharati). Since then, like-minded teachers, students and scientists from various organizations in different parts of the country have been working in close co-ordination with each other at Hoshangabad, participating in the development of teaching material (inexpensive kits and work-books), in the orientation courses for teachers, in monthly follow-up meetings, etc. The enthusiasm generated by this programme is remarkable, and the Government of Madhya Pradesh, which has actively supported this programme from the beginning, now has plans for utilizing on a state-wide basis the approach to the generation of science teaching material and the training of teachers, evolved in the course of this programme.

This example is of considerable interest, not only from the point of view of the participation of academics from several prestigious institutions in a rural-based programme, to bring it to a stage where the state has thought it worthwhile to take it up for wider application, but also from the point of view of the fact that in the initial stages, there were hurdles arising from the fragmentation on the educational scene (e.g. should a university allow its teachers to be absent for several months for working on a mere school-level programme? If so, how is the leave of absence to be treated?) which could be successfully overcome.

These examples are symptomatic of the growing realization by the élite educational, scientific and technological institutions in India that they have to contribute to the solution of problems of the community; and also that there is an intellectual challenge in working on such vital problems of the society. Given the background of these institutions, most of the efforts mentioned relate to the field of education or to probing in the direction of development of "alternative" or "appropriate" technologies—technologies which have an emphasis on utilization of local resources and on employment generation, and therefore a bearing on the reduction of inequality and poverty.[6] There is, however, every reason to believe that, if a political decision is taken and resources are allocated, the scientific community would be in a position to contribute to the solution of the other developmental problems of the society also, including those relating to the satisfaction of minimum needs.

It has to be realized that involvement in such programmes is not an easy decision for a scientist, given his discipline-based training and the value system in which he operates. While his training is usually in a single discipline, most of the real-life problems dealing with human needs are interdisciplinary and involve social action, which the scientist usually shuns. Further, in the academic community in which he operates his peers judge him by his research production in his discipline and his approbation from other scientists and material advancement are also determined by his continued creativity in that discipline. In fact, his peers may think the better of him even if he continues to produce inconsequential but competent papers rather than if he marshals his intellectual abilities and skills to attack some important socially relevant problem. In the latter case, they may think that he has given up his loyalty to science or that he is becoming a drop-out. The scientist has therefore to overcome a sense of insecurity when he steps out of his ivory tower. What is worse, his institution may also harbour a feeling of insecurity if its scientists start embarking on problems which are not supposed to lie strictly within its mandate. Are they not wasting their time? Would such efforts not alter the character of the institution, and dilute its high standards? Would they not make inroads on the already meagre resources available for basic research? Should such applied efforts not really be somebody else's business? In view of this double problem of insecurity—of the individual and of the institution—if the process of scientists' participation in "relevant" activities is to gather momentum, it is important for academic institutions to give up the frequently observed ambivalent stand, and give legitimacy to such "relevant" endeavours by taking a deliberate decision to encourage certain carefully selected socially relevant programmes, where the expertise in the institution can be usefully harnessed. Steps like the creation of the Homi Bhabha Centre for Science Education at TIFR or the ASTRA at Indian Institute of Science, Bangalore, or the formulation of the Karimnagar District Development Project by the CSIR, thus represent an important trend that is most welcome. Once such a decision is taken, and the new programme "adopted", the institution should of course continue to apply its usual high standards of excellence in judging the performance of the staff engaged on the programme, when considering their professional advancement or when making a periodical review of the programme. In that case, there would not only be no dilution of standards, but the new "relevant" programmes will even enrich the intellectual atmosphere in the institution. Moreover, the standards of excellence of the institution will permeate into the other organs of the social system which participate in the execution of these programmes.

With the experience of these examples in mind, I would like to suggest that in order that the scientific community and its research results may be brought to bear on the most urgent problems of society, it would be desirable to evolve a code of conduct for all educational, scientific and technological institutions, which stand for excellence, that a certain identifiable (say 20–30 per cent) component of their research and development programmes should be devoted to socially relevant endeavours. Society may expect a similar commitment from industry and business, too, in the form of a commitment to apprentice training, or adoption of villages or slums or districts for all-round development.

To summarize, in this note I have tried to describe the role that quality institutions can play in attacking socially relevant problems, in particular the problems relating to basic human needs, and the problem of conflicting values which has to be understood and tackled in order to enable them to play this role effectively. Creation of such centres of excellence in the Third World countries, wherever they do not exist, is to be considered as a matter of great urgency, if science and technology are to make an impact on the solution of their basic problems.

Notes

1. See also the background paper prepared by W. K. Chagula for the Pugwash Symposium on the Role of Self-reliance in Alternative Strategies for Development, Dar-es-Salaam, June 1975.
2. The COCOYOC Declaration, adopted by the participants in a UNEP/UNCTAD seminar in October 1974, addresses this problem.
3. H. J. Babha, Remarks at the inauguration of the new buildings of TIFR, 1962.
4. C. Djerassi, "Ronneby 1967 revisited". A paper presented to the 25th Pugwash Conference, Madras, January 1976.
5. Karimnagar District Development—a summary (CSIR, New Delhi, December 1973).
6. A. K. N. Reddy, *Economic and Political Weekly*, Vol. 8, 1109, 1973, and H. N. Sethna, Convocation Address at the University of Roorkee, January 1975.

*Scientific Progress and the Social Goals of Science**

DJERMEN GVISHIANI

Over the past twenty or thirty years a great deal has been written or said about the development of science and its contemporary role and significance. But it is only quite recently that consideration has been given to the problem of deciding what the future course of science should be. In the initial stage of the scientific and technological revolution, a premium was placed on rapid growth in the sheer number of scientific findings and discoveries, on finding new practical applications for these results and on expanding the network of scientific and technological organizations: these were regarded as the "be-all and end-all" of scientific and technical progress.

Only later did it become clear that scientific progress was by definition an ambivalent process and that scientific discoveries could have adverse (and at times, even disastrous) consequences for mankind, the uncontrolled proliferation of arms or the pollution of the environment being but two examples.

The period of the carefree (so to speak) development of science was followed by a period of concern and apprehension, especially visible in the attitude of society towards the implications of the scientific and technological revolution. Society began to think about the aims, directions and rate of scientific development, the ways and means of financing it, and the criteria for assessing the effectiveness of scientific activity.

It was at this juncture that in a number of Western European countries and in America the adversaries of scientific progress came forward and the so-called school of antiscience emerged, rejecting all forms of scientific and technological development.

The antiscience arguments against forced scientific and technological development boil down essentially to the belief that scientific and technological progress necessarily imperils the existence of all mankind. Science introduces new problems into man's everyday life, it is not adapted to his current needs, it is incapable of fostering communication even between distinct branches of the same scientific disciplines. It has, moreover, a negative effect on the development of knowledge in that it gives rise to a mechanistic view of the world and the reductionist identification between

*First published in *Impact of Science on Society*, Vol. 29, No. 3, 1979.

what is real and what is quantitatively measurable. It is claimed that science does nothing to solve fundamental social problems and that scientific research is often geared to military purposes; the high social cost of technological growth is also cited. Such are the basic accusations levelled against science.

The defenders of science rebut these views, holding that science alone is capable of resolving the crisis of contemporary Western society and providing a solution to its major socio-economic problems. Thus, we are now living in a period characterized by a broad diversity of opinions and public attitudes, which may be divided, following Victor Weisskopf's classification, into those of the "pragmatists", who are in favour of a science pursuing only applied and utilitarian aims, those of the antiscientists, who deny altogether the existence of any scientific method and clamour for the throttling of science, and those of "the technocratic optimists" who firmly believe that science can solve all the problems facing mankind.

At the same time, there are a considerable number of clear-thinking scholars in the West who are coming out with suggestions for humanizing science and improving the ways in which its achievements can be used for the welfare and not to the detriment of humanity. Most of these suggestions call for the increased dissemination of information on the findings of scientific research and their potential implications, and for stronger links between scientists and the policymakers responsible for taking decisions about the use of scientific and technological findings; they also raise the question of the role of the scientific community in actively fostering professional standards of conduct among its members.

We are witnessing today a new stage in the evolution of ideas about scientific development, a stage marked, to a large extent, by the more concerted efforts of scientists and the scientific community at large to solve the problems of scientific and technological management policy and to lay down well-defined guidelines for directing scientific development towards the satisfaction of human needs.

It is important to distinguish between two types of development, two courses of scientific and technological progress. One is directed towards the expanding use of natural resources, so as to reduce unit costs without any essential change in extraction and processing techniques (scale effect). The other is directed towards tapping new, less accessible but more extensive reserves of natural wealth, and opens up additional possibilities for future economic development.

Both types of scientific and technical progress are, in fact, interwoven with each other and are often inseparable. It is, none the less, clear that initially the first, "extensive" type of technological progress predominated, whereas insufficient attention was paid to "intensive" probing beneath the surface of nature. This naturally had an effect on the technological culture of the advanced capitalist countries; taking advantage of neocolonial forms of exploitation, they accumulated a large share of the "natural rent" of the developing countries.

The heavy expenditure of raw materials and energy resources in production became the norm in the advanced capitalist countries while technology using raw material and energy resources sparingly was regarded as unprofitable and uncompetitive. Accordingly, there is no point in talking about an era of "cheap" raw materials or "cheap" energy, an era that came to an end with the energy crisis of 1973. This spelled the end of the period when "the world" seemed to have practically limitless economic opportunities and could count on the ready availability of natural resources for industrial use—and, with it, the end of the period when scientific and technological progress was predominantly of the "extensive" type.

We have embarked on a new period, a period when mankind will equip itself with the means to ensure a continual expansion in the supply of natural resources which are potentially available for economic utilization. The transition will require a more searching analysis of the long-term processes taking place in the economic field and will oblige us in our activities to take into account not only current problems but those that may arise in the future. In other words, while there is greater scope for forecasting the implications of decisions being taken at the present time, there is less leeway for making adjustments for current imbalances.

Therefore, though recognizing that there is in principle no limit to scientific and technological progress, we insist that a progressive social economic system should not proceed from that assumption when formulating its practical decisions. Limitlessness is not something that can be decreed. It can become a reality only if, for a determinate level of extraction and processing technology, society realizes that natural resources are limited and is capable of working out a strategy of long-term development ensuring a rational blending of the two types of scientific and technological progress mentioned earlier.

That is why it is becoming increasingly obvious today that a system for managing or controlling scientific development is absolutely essential (based on knowledge of the laws of social development). And it can be said without fear of contradiction that today the requisite basic social conditions are being established which will enable us to meet this challenge.

In this connection I should like to dwell at some length on the salient features of the present stage of scientific and technological development and the directions it is taking. Obviously, the changes we are witnessing are not taking place accidentally but, like any socio-economic process, are the result of a combination of concrete circumstances. While exerting a growing influence on society at each stage of its socio-economic development, the scientific and technical revolution is, in turn, influenced by the particular level of such development. The transition to the new stage is, above all, the result of the interaction of two basic factors.

The transformation of science into a directly productive force—representing a revolutionary change in the infrastructure of social production, in its form and content, and in the character and social division of labour—considerably broadens its social functions.

The contemporary development of science and technology cannot simply be regarded as a sudden transition in the development of the productive forces of society. It is rather a distinct social process brought about by scientific and technological change, a process which not only affects other elements of the social organism (transforming the character of societal relationships and reinforcing the rational, objective elements), but also directly shapes them in a particular fashion.

Scientific and technical development radically alters man's relationship to nature and, most important of all, it makes him aware of his obligation not to violate the ecological balance. This, then, is the first factor which makes it essential that science be managed with lucidity and purposefulness in the interests of progressive social and economic development.

A further reason why the need for controlling scientific and technological progress has been felt with added force in the past ten years is the emergence of problems of global dimensions. One is struck above all by the sheer magnitude of these problems. All are directly or indirectly concerned with the fate of all mankind, a fact which precludes the possibility of their being tackled by a single country or a small number of countries. Wide-scale international co-ordination and action are required. These global problems are also exceedingly complex and are the result of a large number of diverse factors—natural, technological, economic, political, social and cultural. Such problems, therefore, must be studied and solved in an integrated, organic manner, taking into account their interaction with each other; a broad interdisciplinary approach and the combined efforts of representatives of the social, natural and technological sciences are required.

Another important feature of global problems is their acuteness and urgency, aspects compounded by the accelerating pace of all the main scientific, technological and social processes. This necessitates prompt action being taken to ascertain their nature and to take decisions on a theoretical and also on an operational level.

At the present moment two major categories of problems have been identified which have a global and long-range character.

The first category embraces the whole complex of problems concerning the reordering of international relations—staving off the danger of a new world war, halting the arms race, fostering international cooperation on the basis of the peaceful coexistence of states with different social systems, and restructuring international economic relations in an effort, among other things, to overcome the backwardness of the developing countries.

The second category comprises the whole group of complex problems involved in achieving a harmonious interaction between man and nature and satisfying the growing needs of the world's expanding population in respect of food, raw materials, and energy. This category also comprises demographic problems, problems of protecting the environment against pollution, the preservation of the earth's water and atmospheric balance, the control of particularly dangerous or widespread diseases, safeguarding the health of present and future generations against the possible adverse effects of scientific and technological progress, and many others.

Let me stress once again, however, that in the world as it is today, when contacts among peoples and nations are more extensive and varied than ever before and where there are numerous centres of international tension and conflicts, the normalization of international relations is in itself the most important of all global problems and is the key to the solution of all other global problems.

The problem of normalizing the international situation and making the relaxation of international tension an irreversible process is the central issue facing mankind today. Particular importance attaches to halting the arms race. In the event of a major conflict and the unleashing of thermonuclear war, it is useless even to talk about other global problems. There is no more urgent task for the present and the future of all mankind than bringing the arms race to a halt and making the transition to genuine disarmament. Without this, the peoples of the world will suffer further irreplaceable losses in material and spiritual resources. Without this, there can be no realistic solution to the other global problems.

Thus, the transformation of science into a directly productive force in contemporary society, the emergence of global problems requiring urgent solution and a number of other factors point out the need for an effectual and carefully designed scientific and technological policy not only on the national level but on the international level as well.

As already mentioned, a number of basic and essential social conditions already exist which make it possible to tackle the problem of the effective control and management of scientific and technological development.

Among them we would cite, first, the growing social potential and increased internal capacity of science itself as a social institution. The impressive growth of technology, the abundance of qualified scientific manpower, the fact that scientists' efforts are now systematically directed to definite ends, all this gives reason to believe that contemporary science is capable of achievements that were beyond the ken of "classical" science: the simultaneous development of alternative scientific models, or the adoption of different scientific approaches in such a way as to reveal the possible social implications of each and then make it possible to select the one best suited to the humanistic values of our times.

Secondly, there is the fact that mankind now possesses new instruments and the requisite methodology for solving urgent problems, one such instrument of scientific cognition being the systems approach. This approach, which co-ordinates and integrates various types of activities, involving the humanities as well as the natural and technical sciences, seeks to bring to light the maximal number of interconnections and interactions between problems, using both the formal apparatus of logic and mathematics and intuitive, informal methods of thought.

The systems approach to natural and societal phenomena, which focuses on the comprehensive and integrated analysis of the social implications of science and technology and is based on the methodological principles governing complex systems, not only makes it possible to form a complete picture of the interrelationship between the scientific and technological revolution and the development of society but also serves as a practical instrument for guiding the processes of the scientific and technological revolution in the interests of mankind.

An important tool today for understanding and managing the global processes of social development is model building. It frequently serves as a standard method of approach to complex systems which are difficult or impossible to study directly. Model building facilitates the understanding of problems, allowing them to be described as an integral system; it also assists in the unifying of concepts and in working out uniform methods in regard to cognitive procedures.

A major role in integrating the efforts of specialists and in concentrating society's manpower and material resources on solving the crucial problems of the day is played by international scientific research organizations. The International Institute for Applied Systems Analysis (IIASA), founded in 1972 in Vienna, is one such organization. Along with other organizations like the World Health Organization (WHO), the International Atomic Energy Agency (IAEA), the European Organization for Nuclear Research (CERN) and the International Federation of Institutes for Advanced Studies (IFIAS), it contributes to the solution of major interdisciplinary problems, collects information on the scientific achievements of many countries, and strengthens the existing system of international scientific and technological contacts.

Another factor contributing to a more humanistic policy in the field of science and technology is the willingness of the more economically advanced countries to attach less importance to purely economic criteria for assessing the results of scientific and technological activities than was formerly the case. This gradual liberation from the constraints of hard economic rationality is the key to the attainment of other—non-economic—human values; this trend is already observable in the socialist countries, where all technological innovation is first considered from the standpoint of its human implications and only then from that of economic profitability. It is obvious that freed from the obligation of choosing the most highly profitable science and technology, mankind considerably expands the possibilities and applications of scientific and technological development.

Lastly, it is quite clear that the formulation and execution of science and technology policy depend on the fulfilment of certain fundamental scientific, economic and social conditions.

Man is the supreme value and goal of history. In relation to man, science and technology fulfil only an instrumental function. Accordingly, the basic purpose of the management or control of scientific and technological progress must be the creation of conditions conducive to the fulfilment of man's spiritual and physical potentialities.

Today a large number of international organizations and numerous national associations of scientists are advocating a more human orientation of science. Clear evidence of this trend is seen in the decision of the United Nations to hold a conference on science and technology for development, which took place in Vienna in August 1979.

Another striking illustration of this concern is the initiative taken by UNESCO in launching the project on research and human needs. The efforts of the participants in this undertaking are focused on such problems as the development of the systems analysis methodology in relation to the development of science, the study of methodological principles and possible variants for the construction of normative system of values existing in different social groups, cultures and regions, the development of an integrated operational concept of human needs, a comparative analysis of global models from the point of view of human needs, and so forth.

These, then, are the basic factors and conditions at the present stage of scientific development which, in my view, give grounds for believing that a solution will be found to the problems of harnessing scientific and technological progress to the humanistic ideals of our times.

*Science and Intellectual Needs**

IGNACY MALECKI†

Ever since the dawn of history, science has been helping man to achieve his practical goals. At the same time, man's practical needs stimulated new research. Archimedes' law was used to build ships; during the Renaissance, the requirements of navigation led to modern improvements of ancient and Islamic astronomy.

Yet the links between science and practice were virtually random, and progress in science came at the instigation of inventors rather than savants. The steam engine, developed by James Watt in 1782, preceded the description of its thermodynamic cycle by Sadi Carnot in 1824. Applied research began only in the middle of the nineteenth century, directed to the solution of concrete technical problems. Private and governmental, problem-orientated research institutions were then established, with industrial firms organizing their individual laboratories. By the time of the Second World War, vast research programmes were elaborated to achieve military objectives, such as the improvement of radar and the development of the atomic bomb. After the war and until the mid-1960s, the attention of governments and the public pivoted on the economic gains that were forthcoming from scientific research.

The expectations, however, were too optimistic. It was common opinion, for example, that investments in research led to increased profits. By the end of the 1960s, these views were modified, and the voice of criticism could be heard in many countries (especially in the United States). It was found that the return on investment when converting scientific achievement into practice was lower than expected, for example, in the case of research on chemical energy sources. Furthermore, it was realized that increase in a firm's profits was not equivalent to the results of scientific research when applied to the general advantage of society; the necessity of additional equipment for noise control is but one example.

At about this time, two main objections were raised to technical progress based on scientific research:

---

* First published in *Impact of Science on Society*, Vol. 29, No. 3, 1979.

† Academician Ignacy Malecki is Director of the Institute of Fundamental Technological Research of the Polish Academy of Science.

Profits derived from technical progress increase the national income (gross national product) only formally; these profits concentrate, in fact, only within large firms and thus make social disproportions more marked than otherwise.

Industrial activity often leads to socially negative results; for example, to an increase in unemployment and the degradation of man's environment.

These objections can undoubtedly be justified to some extent. There remain, for instance, numerous examples of poverty in countries with the highest *per capita* income and investment in industry has had damaging environmental effects. Let us mention here only the gradual deterioration of Venice and the unbearable living conditions prevailing within the metropolitan area of Tokyo.

But there is no retreat in scientific and technical progress. The only way ahead is to try by further scientific effort to neutralize the negative effects of earlier stages of development.

This state of affairs has led to partial re-orientation of the objectives of research. The goal of transitory profit has been replaced by mankind's needs as the basic objective of directed research. With this broader approach, the problem now consists in establishing links between laboratory research and social goals. But a further step must be taken in order to extend the range of the discussion. We have already commented on the relationship between profit and the material satisfaction of society's needs, but this is not the end of the problem. There also exist non-material needs and non-material interactions and consequences of scientific research that must be taken into account. These non-material aspects must be taken into account as realities in the development of science policy and in estimating the social role of science according to public opinion.

To explain the problem more clearly, let us ask two questions and try to find their answers:

What non-material human needs are there that can be satisfied, directly or indirectly, through scientific research?

What are the social interactions in the development of science and what are their consequences?

When speaking of human needs, one usually means (a) basic material requirements—food, shelter, clothing and appropriate environment; (b) derivative material needs indispensable to the satisfaction of essential needs—water, energy and raw material supplies, transport and medical care; (c) general, but differentiated, needs, for example, the right to work in satisfactory conditions, protection of social rights including those of women, and retirement income.

In satisfying all these needs, scientific research has its part to play. Given the world's rapidly growing population, it is only the results of research that can make it possible to increase agricultural and livestock production, solve the energy crisis, and cope with limited water supplies. The observation also applies to the improvement of our environmental conditions—and this urgent problem is already the object of intensive research.

When considering social needs, it is impossible to disregard the role of the social sciences, in particular sociology and economics, which are capable of finding solutions to growing social problems. But to avoid misunderstanding, it should be clearly stated that scientific research and its results are not capable of satisfying any of the three categories of needs just mentioned. Research and its consequences create, rather, conditions enabling political, organizational, and economic activity to meet these needs. Only in this sense is science a productive force.

There is, in addition, a fourth group of human needs: our intellectual needs, which form an essential element of our overall social needs.

What human beings want is a set of conditions favourable to the intellectual development of both the individual and society as a whole. When the necessary conditions are present, they contribute to growth of creative capacity, ability in logical reasoning, and the acquisition of knowledge.

Intellectual development is necessary not only in scientific and artistic activity, but in all endeavours that create social and economic progress and that occur in organization, technology, politics and economics. In each of these fields, there can exist creative aspirations which are the driving force of social progress. If society, taken as a whole, is deprived of such aspirations, then society is condemned to gradual mental degeneration. Without the human instinct to change or improve reality—by studying the world thoroughly, or by fashioning works of permanent beauty—there would be an intellectually stagnant society, one hardly to be considered ideal even if its level of prosperity were extremely high. The material and intellectual values developed and bequeathed by the preceding generations would inevitably deteriorate, leading to a decline in civilization.

It is obvious that not every individual possesses creative ability and that only part of society contributes actively to the development process. Yet it is important to provide the conditions which enable each individual to participate in the process, while at the same time making it possible for the widest spectrum of society to benefit (at least passively) from the achievements of culture and science.

This is not an abstract conception, because the intellectual demands of man are concrete. They can be reduced to three: (a) for the creation of the psycho-physical predispositions for intellectual development; (b) for the extension and reinforcement of the possibilities of acquiring knowledge; and (c) for the formation of social conditions favourable to, and stimulating, intellectual development.

The human predisposition for intellectual development follows not only from the individual's inherited traits, but also from his material environment and way of living—especially during childhood. It is well known that malnutrition of a baby during its first 18 months results in irreversible changes in the brain, damaging the entire future intellectual life of the human being. This is a problem of vital importance, especially to developing countries: today's hunger may affect the intellectual development of society over a generation or more.

The inherited qualities of an individual have not so far been subject to manipulation. But new possibilities exist, through genetic engineering. Disregarding for the moment the ethical aspects, it can be supposed that such engineering may be capable of producing at least some positive results. We cite, by way of example, the possibility of curing inherited mental deficiency.

Extending the possibilities of acquiring knowledge depends mainly on the nature of the educational system. It is a well-established and reasonable opinion that education is one of the fundamental human needs. But in order to utilize fully the potential intellectual capacities of society, it is insufficient that everyone obtain an elementary education. It is also necessary to open the way to further education for all talented individuals, thereby enabling them to extend their knowledge.

Existing educational systems are not fully adequate to the intellectual requirements of society. This applies both to teaching methods and the selection of subjects to be taught. As a matter of fact, in pedagogics we have not gone beyond traditional empiricism. Furthermore, in view of the growing harvest of knowledge, the proper selection of the subject matter of teaching becomes increasingly difficult, more and more of a responsible task. After all, the teaching process is devoted not only to developing the intellect; it is also committed to the acquisition of knowledge for one's calling in life as well as to development of a philosophy of life.

The most difficult problem, indeed, is how to determine in detail the social conditions that would favour and stimulate intellectual growth. Historical experience has shown that innovative thinking has always encountered, at the beginning, massive opposition. It can hardly be expected that this situation will ever change radically.

Let us now return to the question, "What steps should be taken so that science and the scientific community can contribute to the fulfilment of the needs postulated for society's intellectual development?"

In the matter of intellectual growth, the mechanism of satisfaction is much more complex than in the case of material needs. The problem is not only how to apply the results of scientific research; it is also how to determine what influence scientific activity has on the establishment of conditions favourable to the meeting of intellectual needs. Here it is possible to distinguish a series of mechanisms.

First, the results of scientific research can form the foundation for making non-scientific decisions. For examples, research on the brain and in genetics is the starting point for the rational improvement of man's physiological development and, thus, his intellectual predispositions. Studies in psychology and logic are useful in making the teaching process and research work more effective, while research in the sociology of creativity aims to learn how to elevate the social authority of such work.

Secondly, scientific research serves to improve the mechanisms of reasoning and arriving at conclusions. The turning point in these processes was the advent of computers that search for, collect and process data. Large information systems take over a number of the formal operations once handled by human beings, thereby enhancing the intellect's potential. The development of a technology corresponding to these requirements preoccupies computer scientists, while the rational use of the growing flood of information of all kinds obliges us to use increasingly sophisticated systems of information exchange. Development of these—especially the world scientific and technological information system known as UNISIST—is crucial to international co-operation and the transfer of knowledge.

Thirdly, scientific research is indissolubly linked with teaching at the university level, for two reasons. University teaching is dependent not only on presenting a certain amount of knowledge. In addition, it depends on showing students how to employ scientific methods of thinking; this is of paramount importance to intellectual development. But to teach the scientific method, the instructor himself must be a scientist. Furthermore, if the teacher is to pass on the latest achievements of knowledge it is desirable that he make at least a small contribution to the new knowledge himself and maintain active contacts with the sources of this knowledge. Otherwise, the gap between a given university and the world's scientific centres of excellence will grow and the rank of the university will drop.

Fourthly, the scientific community has a role in the spread of knowledge throughout society, and in stimulating interest in scientific subjects. This is a vital, although often underestimated, task of scientists. The diffusion of science beyond the sphere of institutional knowledge offered by schools has become an important creative factor in culture—one shaping the very intellect of society. And as leisure time increases, the question of how to use this time to best advantage arises. This time is sufficiently long for both rest and the amateur practice of science, as confirmed by the growth of social groups devoted to the study of astronomy, archaeology, wildlife and the like (much as, at one time, the awakening of national sentiment spawned new interest in historical studies).

One should also say that research in fields such as cosmology and molecular biology has stirred the imagination of the public, making the study of these and related problems all the more attractive. Thus, one can say the growth of general interest in science has two consequences: raising the public's general intellectual level, resulting sometimes in the discovery of new talent, and promoting the social prestige of both science and the scientific community.

Fifthly, there is the complex interaction between scientific centres and society. By the time a research centre (institute or university) reaches its "critical mass", it becomes possible for it to conduct widely diversified research. The very atmosphere within the centre becomes such that creative discussion contributes to the enhancement of its research efforts. Simultaneously, the centre exerts social influence on the surrounding community or region, or even the country as a whole.

Finally, a country's intellectual development cannot take place in isolation. Travel abroad, and the growing reputation of a country's scientific community are essential in order to make proper use of the most up-to-date results in every discipline. To gain something, it is necessary to give something; that is, to participate actively in the constant development of mankind's values. Scientific research is so important (although its achievements make fewer headlines than successful sporting events) that it builds for one's country a measure of international respect. Just as artists and research scientists can be the best ambassadors of their countries, scientific achievements can pave the way to mutual technological exchanges. This is why research is often conducted that cannot be applied within the country where the research is done. A team of scientists gaining world renown also brings indirect benefits to a country, independent of the degree of applicability of their work at home. Therefore the criticisms often expressed concerning "prestige research"—research designed to have an effect throughout the scientific world—seem to be unjustified.

I have presented some aspects of the influence of science on society's intellectual development. I have suggested that this influence can take different forms and vary in scope. In the highly developed economies possessing well-established research traditions, science has achieved a stable social position. But significant new discoveries, or the opening of new fields of application, have introduced instabilities, leading to change (as was the case after the first successful flight in space).

In the developing economies, the position of science in public esteem—and in the views of the decision-makers—is still at a formative stage. The stimulation of intellectual development is thus one of the necessities in social policy, and the development in these countries of central and regional scientific institutions is of vital importance. The problems of inspiring interest in science and promoting the prestige of science are thus significant for the whole of a society.

The discussion can be summed up as follows: the application of scientific discovery can provide multilateral economic and social results. In terms of science policy, research orientated towards meeting human needs should be of high priority. Such needs include the intellectual development of society as well as its material and social needs. The state of science in a given country exerts an influence on the intellectual development of its society, and scientific research is helpful in creating those conditions that favour intellectual development, improving the university teaching process, and developing mechanisms to assist in intellectual activity.

Research contributes to the spread of scientific knowledge, developing a creative interest in science, and promoting the prestige of the scientific community within society. A country's international relations in science and the world's recognition of the level of scientific development in a given country contribute indirectly to the overall intellectual development of society. And lastly, the course of society's intellectual development—as well as the methods to stimulate this—depends on local factors; the choice of methods is of special significance in the case of developing countries.

Science and Human Needs: An Artist's Response

BIBI ANDERSSEN*

When I read Federico Mayor's Preface ("Science, Technology and Social Priorities") I was both cheered that such a model should have been put forward, and surprised that it was not seen as completely self-evident. The world today is both wonderful and awful; and if the time has finally come when we can use our desire to live as the mainspring for a new and creative society, it is more than welcome. If we can also make sure that this new society does not become petrified by bureaucracy, it will be more welcome still; with a open-minded attitude, it might just be done. Are we mature enough to create a system based on unselfishness and open-mindedness? Are we brave enough to be constantly open to changes of mind?

If we have to ask ourselves why we are doing scientific research, who will benefit by it, what we mean by evolution, and how to establish social priorities in each country, it can only mean that we have lost contact both with humanity and with our own lives. The answer must be: "for mankind and for all that is alive". That this is not obvious explains why we are on our way to common suicide by nuclear war, starvation, the destruction of natural resources and a deteriorating quality of life. We have got used to this system step by step.

> "Imagination is the strongest power today [says Federico Mayor]. In a world that is dangerously polarized, it is important to have a meeting place for discussion, a role traditionally played by the agencies of the UN. But today, though their role is more important than ever, UN organizations suffer from a certain lack of credibility . . . the organizations have become too big, lose sight of their objectives . . . there is also a 'vanity complex'."

Since most of our organizations share these problems, they must derive from the human inability to continue making individual judgements based on experience but rather to immerse themselves in a machine, become part of it, and ultimately become machines themselves. When something runs out of

*Bibi Anderssen is an actress.

control and man becomes a slave to the machine, then no further communication is possible. Mayor's description of the way things go wrong and how we have lost control of technology is true. His quotation from the words of the Pope is appropriate: "The future of mankind depends on culture . . . peace in mankind's future depends on love."

At the root of all cultures there is a concept involving God and love. When this is lost, civilizations die out. My experience of God is that he is a symbol created by the wisest of men, who have seen the possibilities inherent in human beings, our capacity to grow and become aware of our full resources. The religions which will enjoy a renaissance are those which lead us back from a hierarchy based on power into the capacity every man has to find power within himself. It is clear that these ideas need to be revived, since our societies have bankrupted themselves by our desire to dominate others, in the name of defence. We lack trust in others; we can regain it only when we start trusting ourselves. If we are to establish a system that encourages our own growth and where power means a commitment to the well-being of others, then we will really need both Buddha and Jesus Christ as our teachers. But first of all we have to help ourselves and everybody around us.

"To fall in love is a revolution" says Francesco Alberoni. Do people really change when they fall in love? Of course they do, he says, because falling in love means to overcome obstacles, and the more daunting those obstacles are the more we have the sensation of being in love. We can fall in love with a charismatic leader, or even with an idea; the effect is the same. But the effect does not always last; one day things become "real" and "normal" again and we go back to the old order. "To love your neighbour as yourself" is very difficult for those who live through others and therefore have no self-esteem. Maybe this new age we are entering into will include some ideas about how to fall in love with oneself as well as with society. Marilyn Ferguson describes this process in her book *The Aquarian Conspiracy* as being like coming home after years of sleep, with all the fragments of our personality back in place. When man can find the way to himself, he also discovers a deeper motivation and a new conception of his work. He sees himself as part of the whole, and works with a greater sense of responsibility for what he is doing with his life.

To return to the Pope's words, "that the future of mankind depends on culture". All expressions of man's creativity are culture, but as an artist and actor I can speak only about what is in my own experience. What can an artist do, who cares about his art and the society he lives in? First, he can try to communicate with institutions and leaders, he can write, speak, protest, and try to influence. If that fails, he can organize non-violent demonstrations. If that doesn't work either, if the institutions and the politicians ignore the voices, it leads to destruction—first of the artists and their creations, and finally to violence and war.

Artists are not, as is sometimes said, ahead of their time. They are living in the same society as everybody else, though sometimes they may be sensitive enough to identify and express changes sooner than others.

We live in a dangerous time, because the system has learned to accept the protests with a paternal pat on the shoulder. They listen with only half their minds in gear, then go on as if nothing at all happened. The act of protest may even be absorbed into the bureaucracy itself, as seems to be happening in Sweden. There is a total freedom to speak and be paid for it, which anaesthetizes real protest. Many artists simply fade away in the absence of a challenge; we are not dangerous, we represent no threat, the system absorbs us and we sink slowly into silence. The revolution artists are fighting for today is not militant, but much more subtle: they are fighting for their own sensitivity, for the survival of society's emotional life.

Nijinski, the Russian dancer, once said: "We think too much. We have lost contact with what comes before thought: feeling. To be in contact with our inner selves, both in a poetical, aesthetical and religious sense, is to be in contact with humanity."

The death of a culture comes when artists are faced with a public which has lost touch with its own humanity and has become trapped by its own institutions and laws. Fortunately, not everybody has to be a great artist in the commercial sense of the world. Every human creation is art. All those who can work with their intuition and develop their sensitivity as well as their knowledge and intellect are artists.

What all "human artists" need today is a new kind of courage: the courage to live in a society where their feelings are subjected to disdain and suppression. It is necessary to risk being hurt by the defensive mechanisms of rational minds. It is both dangerous and challenging for those who have amputated their emotions and live only in their minds to be faced with people who take risks, and who dare to accept that the world is not static, and who know that when goals are too firmly fixed society becomes a prison. There is no point in arguing with people who believe they can change the world without changing themselves: their "arguments"—such as they are—will win. Yet openness, sensitivity, and a readiness to change have always been the key to creativity and is today our only weapon against a world that would rather let people starve to death than question its own mechanical logic.

> "We need a type of theatre which not only releases feelings—insights and impulses possible within the particular historical field of human relations in which the action takes place—but employs and encourages those thoughts and feelings which help to transform the field itself." (Bertolt Brecht)

SECTION 2

Methodologies: Practices and Perspectives

Introduction: An International Programme on Research and Human Needs

AUGUSTO FORTI

In the mid-1970s, when UNESCO's programme on Research and Human Needs was first launched, there were at least three major considerations exercising the international scientific community. The first was the acknowledgement that spending on fundamental and applied research had stopped growing in practically every industrialized country as a result of economic difficulties. In many of these countries, emphasis was being switched from fundamental to applied research in the attempt to respond to new problems, such as those raised by the energy crisis. Many scientists thought this trend a dangerous one, with the world likely to face by the end of the century a multitude of new and yet undefined problems, which could be solved only if the "reservoir of knowledge" is filled through investment today in fundamental research.

The second spur to change came from the realization, particularly in developing countries, that very meagre results had been achieved by two decades of UN-inspired application of science and technology to development. It was felt that a new approach was needed in the application of science, tied in with the development of a new international economic order. Science and technology can, in fact, find useful application only when they are rooted in the social, economic and cultural context of a nation or region.

Finally, it was realized that the traditional approach to major problems such as malnutrition had proved inadequate. To face these problems would need an approach transcending disciplinary barriers, and involving co-operation between two or more disciplines.

When in 1973 UNESCO began to investigate the possibility of a programme on research and human needs, it had already had considerable experience in the application of science and technology to development. UNESCO's mandate in the UN family for science, culture and education put it in a favourable position to respond to the requests from the scientific community and member states and initiate a programme, at national and international level, which would look anew at the priorities of research in relation to the world's real problems. Since then, world events have underlined just how appropriate and timely the initiative was. The concept of the new international economic order, promoted by the United Nations in 1974, marked a shift in world priorities; a shift reflected in many countries by a desire for a new development strategy. But the change will not be simply economic. It will include as well a new drive to establish rights and freedoms, a more equitable distribution of resources, a new approach to science and its potential applications, and a new impetus towards meeting human needs such as food, shelter, health, education and the right to meaningful work.

The world's riches are not evenly divided. Some areas are favoured by a good climate for agricultural production, or by the availability of raw materials; others have neither. This alone makes the interdependence of nations inevitable, but with the growth of technology and trade during the post-war period, that interdependence has become so all-encompassing that few nations are any longer in a position to solve on their own the problems posed by the need to apply science and technology to development and welfare.

It was felt that UNESCO, with its broad educational, social, cultural and scientific mandate, could and should undertake the responsibility of making scientific research and its application an integrated ingredient of the new international economic order. The UNESCO member states endorsed this approach during the past two General Conferences, leading among other things to the introduction into UNESCO's programme of a new theme, entitled "Contribution to the determination of research priorities linked with human needs and societal goals". This approach is also in complete accordance with the recommendations of the UN Conference on the application of science and technology to development, which took place in Vienna in 1979.

The main objectives of the programme are the investigation of how research in the social and natural sciences relates to human needs, emphasizing both the promotion of international co-operation and the reorientation of research policy towards the satisfaction of human needs. The purpose, clearly, is to satisfy human needs better, particularly in the developing countries.

The programme has been developed independently in four regions of the world (Europe, Africa, Latin America and Asia) taking into account the wide differences between the different regions. Thus the priorities established for action in Europe are not necessarily the same as those for Latin America or Africa. At the centre, there are two main activities. The first is to try to understand better what human needs are, and the second is monitoring trends in oriented research which could lead to significant improvements in satisfying those needs. An information referral system based on regional centres has been established in order to keep a record of all the activities going on in this field and to provide a link between institutions and individuals working in this area.

To summarize, the main activities of the programme are as follows:

(a) The analysis and definition of human needs in a broad sense, taking into account the variety of social and cultural conditions affecting those needs.

(b) The continuous review and presentation of research developments which could be important to the solution of human needs problems. The review emphasizes the interdisciplinary approach and includes not only material but also the cultural, intellectual and cognitive aspirations of mankind.

(c) The establishment of an Information Service and Data Bases through which the results of activities (a) and (b) can be widely disseminated to decision makers and to the scientific community. The Information Service will disseminate all information and all programmes, projects and research which bear on the human needs question.

(d) The promotion of regional and inter-regional projects on fields of importance to the satisfaction of human needs. The establishment of priority research areas at regional level is also important, such as the inter-regional project on malnutrition.

These activities are supplemented by active training programmes in developing countries, which aim to build up indigenous capabilities for using science in member states to meet human needs. This programme has already achieved some success at the regional level, providing scientists and science policy-makers with useful tools, such as the "Biennial Report on Priorities and Trends in Research related to Human Needs" prepared in co-operation with the International Federation of Institutes for Advanced Study, and a series of monographs under the title "Trends in Scientific Research" (prepared in co-operation with ICSU) as well as a Directory and Bibliography on Research and Human Needs.

One of the targets for the programme has been the subject of malnutrition in early childhood, and a large-scale programme has been undertaken. Other such activities are expected to emerge from the studies of the regional groups.

One of the most remarkable features of the programme so far has been the ready acceptance and co-operation extended by the scientific community. None of our programmes would have been possible without the assistance of national institutions of high scientific reputation. At least fifteen national and international institutions are co-operating with the programme. They include: the International Federation of Institutes of Advanced Study (IFIAS); the National Commission for Informatics, Algeria; the Bariloche Foundation, Argentina; the Science Faculty of the University of Yaounde, Cameroon; the Institut de Recherche d'Informatique et d'Automatique (IRIA), France; the Tata Institute of Fundamental Research, India; the Italian National Research Council; the Polish Institute for Basic Technological Research of the Polish Academy of Sciences; the Science Faculty of the University of Dakar, Senegal; the Centro Latino-Americano de Economia Humana, Uruguay; the Institute for Systems Studies of the USSR; the Instituto

Venezolano de Investigaciones Cientificas; the Institute for Scientific Research, Zaire; the International Council of Scientific Unions (ICSU); the Committee on Science and Technology in Developing Countries (COSTED); and the World Federation of Scientific Workers (WFSW).

These institutions collaborate either within the framework of the regional groups, or are associated with specific working groups—such as that on the Information System. The fact that so many good scientists and eminent institutions are keenly involved in the programme is an encouraging indication of the interest felt by the scientific community.

It may be useful to spell out in rather more detail just how the programme has developed in one region, Latin America, so as to provide a more concrete example of what has been done so far. The programme in Latin America began in 1977, with a meeting organized by the Bariloche Foundation and attended by twenty-eight research workers with a variety of different interests. Those present at this first meeting decided that the Bariloche Foundation would take responsibility for co-ordinating the programme, and a series of recommendations for future action were adopted. One of these was to form a co-ordinated network of persons and institutions in the region interested in working on the subject of human needs. As a first step it was decided to prepare papers giving more detailed study to three topics, which could then be discussed at a seminar. The subjects chosen were: theoretical and empirical studies of human needs in general; human needs in the Latin American context; and prospective studies on the quality of life.

The seminar took place in June 1978 at the headquarters of the Centro Latinoamericano de Economia Humana in Montevideo; forty-three people attended, mostly from Latin America. Some of the papers presented at this meeting have recently been published by the host institution under the title "Research and Human Needs in a Development Strategy for Latin America".

Since then, two further meetings have been held—in Tiradentes, Brazil, in 1979, when the subject was Human Needs and Childhood, and at the Centro de Estudios del Tercer Mundo in Mexico City, in 1980, when the subject was the problems faced by women. On both occasions, there was an opportunity to examine the problems of two underprivileged sectors of society, and how those problems relate to the basic concerns of the Research and Human Needs programme. A book containing the proceedings of these two meetings is to be published in Spanish and English by the Bariloche Foundation and Pergamon Press.

After the Brazil meeting, a permanent consultative board was appointed, composed of people from various Latin American countries. Its objective is to encourage the transformation of what has been until now a series of annual meetings into a permanent research, communication and action network throughout Latin America. After four years of fruitful work and growing influence, the time is ripe for this change. Three projects have been launched to start this new phase: (1) Underprivileged children and women in Latin America; (2) Human Needs research as a key for building up a human development theory applicable in the Latin American context: and (3) Building up a Research and Human Needs reference centre.

One of the basic aims of the programme in Latin America has been to bring to the attention of scientists, policy-makers and the general public the new perspective offered by the Human Needs approach. The term "Human Needs" must be broadly interpreted. Basic needs have the first priority, of course, but the need for identity, participation, affection, creativity, and so on, are also thought of as fundamental. Moreover, they are interwoven with the survival needs, and their satisfaction cannot be denied or postponed without gravely affecting human beings. By taking such a broad view of human needs, the participants in the Latin American programme feel that a proper basis can be laid for a human development model. These concepts of human needs and human development are simple, but far-reaching; they have a remarkable power of unifying those from different disciplines, holding different views, and originating from different regions and backgrounds.

This section deals mainly with the methodological aspects of the programme. There are two principal methodological issues: first, the most appropriate research technique to be used in the field itself, and second, what methodology should be used to determine research priorities.

List of Publications and Documents

"Research and Human Needs—Proceedings of the Moscow meeting on the Unesco project" (Institute for Systems Studies, Moscow 119034, USSR) (Russian only).

"Colloques IRIA—textes des communications: Problèmes posés par la modélisation mathématique des phénomènes sociaux" ("Problems raised by mathematical modelling of social phenomena") (English/French, 26/27 February 1979) (Institut national de Recherche en Informatique et en Automatique, B.P. 105, 78150 Le Chesnay, France).

"Investigación y necesidades humanas en una estrategia de desarrollo para américa latina" (Centro latinoamericano de economia humana, Certito 475, ler piso, Montevideo, Uruguay).

Impact of Science on Society, Vol. 29, No. 3, July–September 1979. "Research and Social Goals—Science in the Service of Mankind" (UNESCO Bookshop, UNESCO, Place Fontenoy, 75700 Paris, France).

"A Directory and Bibliography on the Theme Research and Human Needs", 1st and 2nd editions (UNESCO, SC/SER, Place Fontenoy, 75700 Paris, France).

"Biennial Report on Research Trends and Priorities in Relation to Human Needs Problems", September 1978 (UNESCO, SC/SER, Place Fontenoy, 75700 Paris, France).

"Second Biennial Report on Research Trends and Priorities in Relation to Human Needs Problems", October 1981 (UNESCO, SC/SER, Place Fontenoy, 75700 Paris, France).

The Role of Methodology in the Research and Human Needs Programme

A. FORTI AND O. ZAKHAROVA

The problem of redirecting scientific research towards the satisfaction of the most urgent human needs represents an interdisciplinary task very typical of our time. But scientific scrutiny of this new area is impossible without a comprehensive methodological approach which would allow a variety of scientific disciplines to be brought to bear on the problem. That is why it is important to elaborate adequate methodological tools and approaches so that this complicated problem can be attacked. The main methodological difficulty comes from the fact that the sociological and methodological aspects of scientific development, on the one hand, and human needs, on the other, have been traditionally studied within different scientific disciplines. Even within one of them, sociology, there are two almost independent branches dealing with the sociological aspects of human needs and scientific development.

Work on the Project necessitates the elaboration of a comprehensive methodological approach which would take into account both aspects of this complex problem. This is why methodological problems were one of the first preoccupations of those who have initiated the UNESCO Project on Research and Human Needs. The first meeting of the methodological group created within the framework of the programme was organized in Moscow following the decision of the Paris conference of experts (October 1977). It stressed the importance of adequate methodology in the realization of the project, the application of the most effective and up-to-date methods, and interdisciplinary approaches involving close co-operation between scientists representing different disciplines and various countries.

Having stressed the importance of close international co-operation in this field, the methodological group of the Project agreed on the following basic principles:

1. Human needs, being primarily a socio-cultural phenomenon, reflect the specific social and cultural context of people's lives. From this comes the need for regional approaches to the determination of urgent human needs within a specific socio-cultural climate and depending on the level of economic development. Human needs depend on systems of

values which have developed during previous experience and depend on the conditions of man's existence in any specific region. Which needs are most important depends on these systems of values; hence the importance of studying them within the framework of the Project.

2. Human needs constitute a very dynamic, continuously changing system, characterized by the continuous appearance of new needs and values and by the displacement of the boundary between the "minimal" and "non-minimal" levels of their satisfaction. Forecasting the way this system will develop and change turns out to be a task of primary importance. It is, therefore, especially urgent to study the values of various social groups, on the one hand, and the changes in the position of these groups within the social structure, on the other, and proceeding from this, to forecast the changes in the corresponding systems of values.
3. The key element in the formation and the functioning of the system of human needs is the process of creative social activity, man's habit of changing and transforming his environment in accordance with certain systems of values. The satisfaction of man's need for creative and meaningful work is not only a necessary prerequisite for satisfaction of many other needs and for the harmonious combination of material and spiritual needs, but also opposes the development of "pseudoneeds" and overconsumption.
4. The problem of satisfaction of both spiritual and material human needs represents one of the most important global problems of our time and demands for its solution close international co-operation, especially within the world scientific community, together with exchange of information. This necessity comes from the fact that scientific knowledge is universal and international, and should be applied everywhere in order to make the most rational use of resources spent on science worldwide.
5. The global problem of a fuller satisfaction of human needs requires, for its solution, a redirection of research and its humanization.
6. This reorientation cannot be achieved without first establishing an optimum relationship between fundamental and applied research and development, improvement in R and D policies, and mobilization of public opinion as a whole and the scientific community in particular.
7. An interdisciplinary systems approach is necessary which should take into account the dynamic, active nature of the problem.
8. The functioning and development of science as a subsystem of society depends not only on the goals which society puts before it, but also on the system of values of the scientists themselves, which is therefore an indispensable element in achieving the aims of the project.
9. Research—and the assessment of human needs in their relation to

research—is a part of a combined analysis of the problems of world development. The kind of research needed to satisfy people's needs differs from one country to another. The methodology of socio-economic indicators should be used to classify countries and regions according to their type and level of development. The immediate task is to work out indicators which will make it possible to describe how resources can be formed and distributed for consumption.

10. There is close interrelation between research on human needs and the modelling of national, regional, and global development. On the one hand, research on the problem of human needs provides a scientific basis for determining the goals and criteria of long-term development at regional and global levels. In particular, this research can yield the information necessary to develop and to formalize those criteria and to build normative and optimization models of global development. On the other hand, research using mathematical simulation can give an idea of how likely it is that certain human needs can be satisfied in different regions, and some indication of the necessary measures, resources, time periods, and so on. To reach useful results it will be necessary to construct mathematical models at different levels of aggregation from national to global. It is considered sensible to carry out of the modelling *in national research centres* under the overall guidance of UNESCO.
11. The project Research and Human Needs is complex and includes qualitative aspects difficult to formalize. Therefore, some members of the methodological group expressed the opinion that the elaboration of specific systems analytical methods with mixed qualitative and quantitative aspects is a very important task.

Working from these principles, the methodological group agreed on the necessity to focus its activity on the following main tasks:

(1) *Development of a conceptual model of human needs applicable to the aim of the project,* which consists of research on:
—systems of values in different social groups, cultures, and regions in different spheres including labour, leisure and family life, the mechanism of the interrelation of physical and spiritual needs, methodological principles for development of normative systems of values, including those existing within the scientific community.

(2) *Comparative analysis of global models from the point of view of the main aim of the project.*

The important problem here consists in developing global modelling methods and constructing scenarios which will permit simulation of world development alternatives, taking into account parameters connected with the project. A simulation system set up this way to comprise formalized as well as non-formalized elements should permit research workers to search for solutions to global problems which will assure the optimum level of satisfaction of needs of the widest masses of the population.

(3) *A sociological survey of the opinion of the public in general and of the scientific community in particular on the problems and perspectives of the research.*

The results of a thoroughly-prepared sociological inquest conducted on behalf of UNESCO would permit the creation of a necessary foundation for a more concise definition of the further efforts to develop the project.

Studies of the role of scientists and their organization in the humanization of research and its re-orientation towards the satisfaction of human needs, should form an indispensable part of this survey. It would concentrate on changes in the attitudes of the scientific community, and a study of the mechanisms affecting these processes, as well as the mechanism whereby scientists affect public opinion, the elaboration of research policy and its re-orientation in accordance with human needs, and, finally, a study of the systems of values existing in scientific community. Some parts of this investigation could be conducted in co-operation with the UNESCO programme Science in the Contemporary World.

(4) *Development of the systems methods of analysis and evaluation of research problems and fields on the basis of criteria related to the satisfaction of human needs.*

One of the possible research directions to develop this system is the analysis of a problem within the framework of its three components: research, fields of activity, and needs.

(5) *In addition to these research activities, the working groups also consider it important to conduct research on the following:*

—General methodological aspects of the relationship between the "external" influences on research priorities and the internal logic of its development.

—The role an interdisciplinary systems approach can play in the elaboration of an integrated notion of needs.

—Comparative analysis of techniques for the collection and generalization of information on the problems of human needs at regional and national levels.

—Verification of the mechanisms by which research priorities are set and of channels through which these mechanisms are effected.

—Review, analysis, and synthesis of the results of relevant socio-economic studies, both in the main fields of world science and in the scientific departments of UNESCO. It was proposed that an information centre be created in conjunction with the project, which will in future be transformed into a data bank and an information retrieval system to issue periodically the revised reports for countries and regions.

—The working out of proposals on the elaboraion of a system of socio-economic indicators of development relating to the project and of their evaluation based on the actual statistics.

The Methodological Group considers that the scheme for evaluating scientific research requires the writing of specialized software. Software could be written by using modular programming techniques in order to make it easily adaptive to the specific needs of each user. In this regard, the members of the methodological group have mentioned a software of this kind, oriented towards macro-economic modelization (MODULECO) and being developed by a club of French institutions under the Research Institute for Informatics and Automatics (IRIA), and a group specialising in systems modelling of global development at the Institute for Systems Studies (USSR), both of which could collaborate with the project.

Research Priorities and Human Needs

CARLOS MALLMAN*

It is well known that the health of human beings can be better assessed if they are examined as a whole by one physician and simultaneously as a set of parts by different, specialized, physicians. In this way, "holistic" and "analytical" forms of understanding can be combined. Each kind of understanding is necessary but not sufficient.

In the same way, human beings can be understood still better if seen in their social context. This is why in a programme on research and human needs the research priorities are necessarily related to the generation of holistic knowledge. Analytical knowledge is not excluded, of course; both kinds are necessary, but more emphasis must be placed on holistic knowledge to make up for the fact that in recent centuries the dominant trend has been towards generating analytical knowledge. UNESCO is the only organization of the UN system whose mandate is related to the generation, maintenance, use, distribution and interchange of knowledge. As a consequence, it is the UN organization which should promote the study and solution of this problem.

The first requirement is an attempt to list and classify human needs. What follows (Table 1) is a very rough first approximation to such a list. In it, human needs are grouped according to categories of factors which satisfy them, which I shall call "satisfactors". Three examples may help to clarify this concept.

I may, for instance have a desire to eat a beef steak and, if I do so, I shall have momentarily satisfied my need of nourishment. I would like to participate in my family's decisions on how to spend the monthly income and, if I do so, I shall have temporarily satisfied my need of participation. I would like to move back to my home town in order to satisfy my need for roots. In these three examples, the steak, the family decision-making and the home town are the respective "satisfactors".

From this list of needs it is possible to begin to construct a set of theoretical human rights and responsibilities which correlate with the needs. The purpose of providing such a list is to try to provide some idea of which directions we should be moving in the attempt to improve the quality of life of the world's population.

---

* Professor Carlos Mallman is Chairman of the Bariloche Foundation in San Carlos de Bariloche, Argentina.

TABLE 1

Classification of needs according to categories of needs			Classification of needs according to categories of satisfactors which mainly satisfy them	Personal: A) Psychosomatic or Intra-human	Extra-personal: B) Psychosocial or Inter-human	Extra-personal: C) Psychoecological or Extra-human
		III) Subsistence	I) Maintenance	a) Nutrition, Rest, Exercise	b) Earning-work, Reproduction, Social habitability	c) Shelter, Clothing, Physical habitability
	VI) Existence	*V) Security*	II) Predictability	a) Prevention, Cure, Defence	b) Prevention, Restitution, Defence	c) Prevention, Restitution, Defence
Living			IV) Love	a) Belief in oneself, Self-love, Identity	b) Friendship, Sexual and Family love	c) Rooting Attachment
		VIII) Belongingness	VII) Understanding	a) Psycholization, Introspection, Study	b) Socialization, Education, Information, Observation	c) Habitatization, Observation
	XI) Co-existence	*X) Dignity*	IX) Participation	a) Liberty, Independence, Autonomy	b) Autonomous participation in decision	c) Autonomous participation in management
	XVI) Growth	*XIII) Development*	XII) Variety	a) Self-recreation	b) Social recreation	c) Recreation in the habitat
		XV) Renewal; *XVIII) Transcendence*	XIV) Creation	a) Creation by oneself	b) Creation of social environments	c) Creation of habitational environments
Actualization			XVII) Meaning	a) Self-realization	b) Historic, Prospective and Religious Meaning	c) Weltanschauung
	XXI) Perfection	*XX) Maturity*	XIX) Synergy	a) Authenticity, Equanimity, Security, Humility	b) Solidarity, Justice, Altruism, Generosity,	c) Beauty, Ecological Equilibrium

1. *Maintenance*

Children, elderly and sick people have the right to be maintained, and the other inhabitants of the planet have the personal and joint responsibility of maintaining them and maintaining themselves. This implies:

(a) From the *personal* point of view, having adequate nutrition (air, water, food); rest (sleep, leisure); and exercise.

(b) From the *social* point of view, to earn adequate income for the work they do and to have an adequate social environment (free from personal and social violence) and to have the right to reproduce (the generation, as a minimum, of one human being per person).

(c) To have adequate clothing and shelter provided with the basic sanitary and energy services.

2. *Predictability*

Children, elderly and sick people have the right to be protected and the other inhabitants of the planet have the personal and joint responsibility of protecting them and protecting themselves. This implies:

(a) From the *personal* point of view, to promote the use of preventive health techniques (vaccinations, food stocks, check-ups, etc.); to treat illnesses as soon as the symptoms are detected (rapid access to health services); and to defend oneself and others by all available legal means.

(b) From the *social* point of view, to promote the use of techniques for the prevention of social disasters, aggressions, torture, repression, war, pillage, mass disorders, etc.); restore to their original conditions the people affected by this kind of disaster; and defend the people affected during and after the disaster.

(c) From the *habitational* point of view, to promote the use of techniques for the prevention or avoidance of physical disasters (earthquakes, hurricanes, floods, fires, etc.); restore to their original conditions the people affected by this kind of disaster, and defend them during and after the disasters.

3. *Subsistence*

To *subsist* one has to have maintenance and protection, so that the rights and responsibilities (1) and (2) apply.

4. *Love*

Children, elderly and sick people have the right to love and to be loved and the other inhabitants of the planet have the personal and joint responsibility of loving them and loving each other. This implies:

(a) From the *personal* point of view, to believe in our intrinsic potentialities of existing, co-existing, growing and perfecting and to manifest this belief by developing them.
(b) From the *social* point of view, to have the right to live intimately with one other person at a time; love, respect, grow and help to make perfect our fellow human beings.
(c) From the *habitational* point of view, to love, respect, take care of and make good use of our land, flora, fauna and settlements.

5. *Security*

To feel secure one has to feel protected and loved. Consequently the rights and responsibilities (2) and (4) can be used.

6. *Existence*

To exist one has to subsist and have security. This means to have maintenance, protection and love. Rights and responsibilities (1), (2) and (4) apply in this case.

7. *Understanding*

The children of the world have the right and responsibility of acquiring knowledge and information through formal and informal means. The other inhabitants of the planet must have free and equal access to all knowledge and information available. This implies:
(a) From the *personal* point of view, to understand oneself physiologically and psychologically.
(b) From the *social* point of view, to understand other human beings, our relations with them and with society.
(c) From the *habitational* point of view, to understand nature and human settlements and their relationships with him, with other human beings and with society.

8. *Belonging*

To belong one has to feel loved and understood and must love and understand. As a consequence, the rights and responsibilities are given in (6) and (7) apply.

9. *Self-reliance*

Children should be brought up in environments in which participation in decisions is the norm, and they should be progressively incorporated into the participation process. All other human beings of the planet have the right and responsibility of participating directly, when the type of decision so allows, or if not, indirectly, in all the decisions which affect them. For this purpose, every person should enjoy the maximum liberty, independence and autonomy which is possible without infringing the same rights belonging to his fellow human beings. This implies:

(a) From the *personal* point of view, to govern oneself, which means to have one's own points of view and emotions and to know how to sustain and defend them.

(b) From the *social* point of view, to participate in the decisions which affect human beings and to express oneself freely and responsibly.

(c) From the *habitational* point of view, to participate in decisions which affect nature and settlements.

10. *Esteem*

In order to have esteem, it is evident that one has to understand and govern. As a consequence, the rights and responsibilities spelled out in (7) and (9) apply.

11. *Co-existence*

To co-exist or live together, there must be belongingness and esteem which means having love, understanding and self-reliance. Rights and responsibilities (4), (7) and (9) apply.

12. *Variety*

All human beings in the world have the right, during a certain minimum time per day, month and year, to renew themselves through recreation. This implies:

(a) From the *personal* point of view, to know how to enjoy works of art, literature, humour, scientific results, music, sports events, etc.

(b) From the *social* point of view, to know how to enjoy together with others the items mentioned under (a) and (c) and very specially, profound personal and group communication with other human beings.

(c) From the *habitational* point of view, to know how to enjoy nature, human settlements, architectural masterpieces, parks, sports, etc.

13. *Development*

In order to develop, one has to participate and recreate. Rights and responsibilities (9) and (12) apply.

14. *Creation*

All human beings of the world have the right to develop their creativity. This implies:

(a) From the *personal* point of view, to produce creative work.
(b) From the *social* point of view, to improve the social environment.
(c) From the *habitational* point of view, to generate or help generate better physical environments.

15. *Renewal*

In order to renew, one has to recreate and create. The rights and responsibilities given under (12) and (14) apply.

16. *Growth*

To grow, one has to develop and renew, which means being self-reliant, recreative and creative. Rights and responsibilities given under (9), (12) and (14) apply.

17. *Meaning*

Every human being in the world has the right to choose freely the meaning which he wants to give his life, his society and his habitat. This implies:

(a) From the *personal* point of view, to discover and realize what he was born to be. This will give meaning to the relation with himself.
(b) From the *social* point of view, to define and realize the meaning he wishes to give to his relation with the past (history), with the future (prospective) and with the human species (religion). This will allow him to give himself social meaning.
(c) From the *habitational* point of view, to define and realize his own perspective on the world, which will give meaning to his relation with nature and human settlements.

18. *Transcend*

In order to transcend, one has to create and give meaning. The rights and responsibilities given under (14) and (17) apply.

19. *Synergy*

Every human being in the world has the right and responsibility to increase his three synergies and those of his fellow human beings. This implies:

(a) From the *personal* point of view, to tend to be progressively more authentic, equable, secure and humble.
(b) From the *social* point of view, to tend to be progressively more neighbourly, trustworthy, just, altruistic, generous, responsible and adaptable.

(c) From the *habitational* point of view, to act on nature and human settlements in such a way as to improve their beauty and ecological equilibrium.

20. *Maturity*

In order to mature, one had to give meaning and have synergy. Rights and responsibilities (18) and (19) apply.

21. *Perfection*

In order to be perfect, one has to be mature and to transcend, which means to create, to give meaning and to be synergic. Rights and responsibilities given under (14), (17) and (19) apply.

Research Priorities

We now come to the topic which is the purpose of this article. First, a general statement: Research priorities should be directed towards the quickest possible attainment of the sustainable satisfaction of human

TABLE 2

Existence needs	Problems	Effects	Areas
Maintenance	Maldistribution of income-work, Physiological and Psychological Malnutrition, Population explosion, Lack of shelter	Starvation, misery, lifelong impairments	Mainly in less developed countries and in discriminated sectors of developed countries
Predictability	Armaments race	Danger of total destruction, less security, misuse of resources	Mainly in developed countries and as a consequence also in less developed countries
Love	Patriarchal and authoritarian personal and social structures and functioning	Non-authentic dependent love relations, discrimination of women, authoritarianism sickness in men	In all countries but less in subsistence economy areas
Love	Outward and materialistic personal attitudes	Alienation from oneself, lack of self-love and self-belief	Mainly in Western countries
Love	Depredation of nature, exploitative relation with nature	Ecological catastrophes, pollution, lack of love for nature	Mainly in developed countries and in developed areas of less developed countries

existence needs (maintenance, protection and love), according to the desires expressed by the population itself and in such a way as to contribute, at the same time, to the increased satisfaction of the other human needs (co-existence, growth and perfection).

The first goal should be the eradication of the production of new, irreversible injustices in existence needs. In addition, greater priority should be given to research programmes when the number of people to be helped by them is large, and when the life expectancy and security of the people are low.

Table 2 gives a list of problems described in a very aggregated way. The problems mentioned in the table should certainly have the highest priority in the Research and Human Needs programme.

A Human Needs Approach to Science

OSCAR NUDLER

The UNESCO programme on **research and human needs** focuses on the link between human needs and scientific research. The problem is not new. From Francis Bacon who, on the eve of the scientific revolution, expressed his unlimited confidence in the power of science to satisfy human needs, up to the contemporary critics who see in the expansion and increasing power of science and technology over human life a grave threat to the future of humanity, there is a whole range of different answers to the question about how science and human needs are linked. But it is not my purpose here to add to a debate framed in these terms. I would like to take a step aside and rethink the key concepts involved in our problem: human needs and scientific research. It is my conviction that there is a route by which scientific research and the satisfaction of human needs can be made to converge. But I am also sure that if the present course continues to be followed, the gloomy forecasts of the critics of science and technology might very likely come true.

This is not the place to develop a fully-fledged theory of human needs. I shall outline only some of the most basic requirements which a satisfactory human needs theory should, in my view, be able to meet. The first requirement of an acceptable theory is to take into account the *whole range* of human needs and not only some of them. Nutrition, sex, love, security, knowledge, freedom, participation, identity, meaningfulness, are among the more fundamental needs, that is to say, those which define the human condition. What frequently happens is that some needs—for instance, sex in the Victorian society, or nutrition in a poor village in the Third World today, or national identity for an uprooted people—are inadequately satisfied so grow in apparent importance to monopolise the picture. But a hungry man is still a complete man. His first priority may be to eat, but his other needs do not disappear for that reason; they are merely waiting their opportunity to claim attention. Very often theories about human needs do not take due account of this fact, reducing the human being to very few dimensions. Economic theories illustrate this attitude. But although it is perfectly legitimate for political action to set priorities, valuing some needs as more urgent than other, this does not mean that a comprehensive view of human needs should not illuminate and guide such action from the very beginning. Otherwise any progress achieved in the satisfaction of certain human needs may be vitiated by a feeling of frustration caused by the neglect of other needs. This is well exemplified by some societies which have been able to satisfy the material needs of their population, but in which the problems of increasing violence, drug addiction and suicide are indicators of a profound neglect of some other fundamental human needs.

This first requirement leads us to a second one: human needs *make up a system.* This view implies that the extent to which a given need is satisfied also has consequences on other needs. Traditional theories about human behaviour attempted to explain this by proposing lists of instincts, hypothetical entities isolated from one another. Subsequently instincts were substituted by *needs,* without the atomism of instinct theorists being altered. The different theories which were proposed were essentially different lists of needs. If the study of behaviour revealed that some aspect of it could not be understood within the categories of needs postulated until then, a new need was simply added to the list. Human needs as a *system,* with its own internal structure and dynamics, was not set out as a problem. An alternative approach has emerged since authors like Abraham Maslow began to look for the interconnections between human needs. Today, the study of the human needs system is a legitimate and fascinating field of research.

My third requirement is that an acceptable needs theory must embody the human needs system as an *open* system. As is well known, Ludwig von Bertalanffy introduced the concept of open systems to describe those systems which are characterized by an exchange of matter-energy with their environment. Applied first to organisms, the concept was extended by von Bertalanffy himself to human beings although in this case the exchange is not purely physical, but also mental and to a great extent, symbolic. The individual is thus conceived of as a system open to its natural and, especially, to its cultural, environment. It is in fact an *extremely* open system since its constitution and development are inconceivable outside its culture. The human needs system is certainly not immune to cultural conditioning. Any theory which does not take this properly into account—and that includes very influential theories such as the Freudian one, for example—cannot be wholly satisfactory. At the same time, however, it would be wrong to assume—as social determinism does—that human needs beyond the biological ones are no more than a product of culture. To take only one example, a small child who is well-fed and physically cared for but who does not receive affection is likely to suffer serious psychic and physical disorders. Studies by Spitz, Bowlby and many other child psychologists have documented this abundantly and beyond all doubt.

To summarize, then, a satisfactory theory of human needs should consider the human being as an open system, though not to the point of taking away all his autonomy and making him a puppet of the culture in which he is immersed. To express it briefly, a satisfactory theory of human needs must be sensitive to the subtle dialectic between the human condition and culture and must avoid over-emphasizing either of the two sides.

A human needs theory built on these lines is likely to have a big impact on several fields. One of the most crucial of these is the development field where such a theory could provide the foundation for a new vision which I call the *human development approach*. To outline very briefly this human-centred approach to development, let us follow the consequences of adopting it. Consider, to begin with, **economic development.** From a human development perspective, the goals to be attained undergo radical changes. Economic development ceases to be an end in itself, and becomes a means for human enhancement. Interest centres on achieving a balanced, self-sustained economic development, even at a slow pace, rather than rapid economic growth at the cost of increasing external dependence and internal inequalities and inequities. Economic austerity is not necessarily seen as a bad thing. There is no urgent desire to match the economic standards of developed countries, as in the GNP approach. There is, instead, urgency in eradicating the material *and* non-material poverty and deprivation of all sectors of the population. The need for economic growth is recognized and dealt with, but in the framework of a concern for *all* social and human needs.

Turning now to the *political* aspects of development, the human development approach entails an entirely new vision of political change. It regards as unacceptable the view that an enlightened élite should be responsible for political change. When political action is seen as one of the ways which the entire population can begin to satisfy its needs, then democratic participation cannot be denied or even postponed. Participation is not then only a *goal* but also as a *means* of political progress. Of course, the political structure of a large modern society is quite complex and does not allow the simple participatory techniques suitable for a small community. But this does not justify the concentration of power in a small, usually closed, group. It implies only that a great deal of political imagination and social engineering is required. One of our biggest problems today is that most of our political ideologies come from the eighteenth and nineteenth centuries and have not been re-shaped for our world. My conviction is that a human needs philosophy could contribute to this re-shaping or even to the emergence of a fresh political vision, able to guide us towards increasingly humanistic forms of political organization, far removed from Orwellian or Huxlean nightmares.

The Bariloche world model, for example, showed the feasibility of a concrete application of this basic needs approach. Nevertheless, although the satisfaction of these basic or "first floor" needs is no doubt a fundamental aim of social development, the satisfaction of the "second floor" needs should not be left to chance.

If the human needs and human development philosophy were consistently adopted, a changed view of scientific research would follow. Since other contributions to this book deal extensively with the human needs approach to scientific research, I shall refer only briefly to it. Let me enumerate some of the points which follow from this approach. One immediate consequence is that the results of research are to be applied for enhancing human ends, that is, human life, growth and happiness.

But it is not only the *uses* of science that are affected. A point not always recognized and which I would like to stress is that the scientific enterprise itself is bound to undergo profound changes under the impact of a human needs philosophy. I could mention in this connection the problem of human relationships *within* scientific institutions. In spite of individual efforts, a non-humanistic attitude is more widespread than is desirable. Institutional science is more often than not highly competitive and authoritarian. A human needs approach to scientific research calls for a replacement of these negative values by their very opposites, that is, co-operation and solidarity. Scientific communities which live up to these values would develop an inner commitment to contribute to the satisfaction of the larger communities which they serve. It would be unwise, however, to think that reaching such a commitment and acting according to it is easy. Very frequently external pressures push in exactly the opposite direction. But some steps could always be taken. Scientists have a specially great moral responsibility toward society and cannot abdicate from it under any circumstances.

Together with the transformation of human relationships in scientific communities which would take place when a human needs approach is adopted, a reinforcement of the will to break rigid disciplinary and academic boundaries is also likely to occur. This does not imply simply erasing disciplinary differences and specializations or mixing up different disciplines. Each science and particularly each great scientific theory or "paradigm"—to use Kuhn's popular term—is a way of looking at some specific aspect of reality. More precisely, each theory involves the construction of an abstract object which is supposed to make contact at some definite points with one sector of the real world. This approach to reality is an essential characteristic of the scientific way; to abolish it would be tantamount to abolishing science itself. But even if this is granted, integration among disciplines still appears to be a necessary step for achieving scientific understanding of the world. Integration, as I understand it, implies at least two inter-related processes. First, setting out and facing problems irrespective of disciplinary rigidities. Second, taking into account, in the very stage of theory construction, the knowledge provided by other theories and disciplines which look at the same or at related aspects of reality from a different perspective, with a different purpose.

We have also the problem of selecting research priorities. According to the presently prevailing philosophy, emphasis is placed on stimulating research which can be applied to increase the economic and political power of the State and/or the big corporations. Human needs are at best taken into account when their satisfaction is compatible with those ends. But when there is conflict between the desires of the powerful and those of the population, the latter are usually sacrificed. (The best example of this is, of course, the arms race.) The human needs approach would give priority to research targets linked to human needs, but at the same time should not involve a dull pragmatism hostile to abstract research and theorizing. As is well known from the history of science, pure research has very often had totally unexpected and far-reaching applications. The human needs criterion should be a policy guideline to be followed with intelligence and respect for the scientific endeavour, not used to destroy the scientist's freedom and hence his creativity.

I hope that a human needs approach will eventually prevail in scientific research. But let me end with a word of caution. I do not think that science, even if it followed an approach like the one suggested here, could be a panacea for all social problems. Science, being an essentially intellectual and rational enterprise, cannot satisfy some of the more fundamental human needs. Scientific utopianism is as mistaken as other forms of utopianism which do not pay enough attention to the complexities of human society and the human condition. I am convinced, however, that a science grounded on a human needs approach could make a substantially greater contribution to facing human and social problems than the one science makes today.

Assessing the Contribution of Research to Human Needs: a Possible Methodology

S. A. PETROVSKY *

Because science originated as part of man's practical activities, it has always served to satisfy human needs. A mechanism has always existed to ensure a more or less satisfactory agreement between needs and the direction in which science develops. But this mechanism has no fixed character; it links two things, science and human needs, which are themselves subject to change, and exists within a changing social milieu, so it too has been subject to modification. This evolution continues, although today attempts are being made to direct it more deliberately.

Such attempts reflect the realities of modern life. Today even the most elementary of human needs are not being met for a considerable part of the world's population and in the future things may get worse (failure to solve social problems, the shortage of fuel and raw materials, uncontrolled population growth) creating conditions under which it will be ever more necessary to direct science towards the needs of man. But at the same time, solving this problem by traditional means is becoming more and more difficult. The rapid growth of science, the range of research projects, and the proliferation of scientific information are matched by an equally rapid development in the needs felt by people. At the same time, science can seldom provide an immediate result in satisfying people's needs; as a rule, much more work is necessary before the results of research can have their effect.

This situation is a familiar one for scientists: a system whose complexity constantly increases eventually defies conventional analytical methods. I believe that such complex problems can be tackled only by combining traditional analysis with the modern method of the systems approach.

As I have already indicated, a substantial amount of activity must take place before scientific results can be turned into a means of satisfying needs. Scientific results are applied in a variety of ways, producing goods or services, which in turn may satisfy needs. So it is inappropriate to think simply in terms of research and needs; one ought instead to think in terms of research—applications—needs. Then the problem divides naturally into three parts; estimating how important any research result may be in terms of its potential applications; estimating the importance of such applications in

* Dr. S. A. Petrovsky is at the International Research Institute for Management Studies, USSR.

satisfying needs; and finally, assessing the importance of the needs themselves.

One cannot make estimates of this type without first defining exactly what is meant by the concepts involved, and by being constantly aware of the context of the problem being analysed. For example, the need for pure air to breathe and food to eat might be rated as needs of equal importance, since life is impossible without either. But such a comparison is irrevelant in the context of human needs if—as is the case—the supply of pure air presents no problem. There is therefore no point in giving this need great prominence or including it in a list of human needs. (I am not, of course, considering here the question of air pollution in modern industrial countries, where the need for pure air is a serious problem.)

When we consider the relative importance of research fields in satisfying human needs, it is obvious that what we are doing is assessing their ability to solve problems. Through productive activity (using productive in its widest sense) a set of human needs is satisfied. The needs are those overriding problems to which all other activities of the society are, in the final analysis, subordinated. In this sense, research is a way of stepping up the efficiency with which a society can solve its problems.

From this perspective it is possible to assess the importance to be attached to each element of the problem. Society itself provides a measure of which problems it perceives to be the greatest by the way it distributes its primary resource, labour. Thus the relative importance of different fields of activity with respect to any given need can be measured by the proportion of the labour force engaged in that activity whose task it is to solve that particular need.

Similarly, the importance of different research projects to any given field of activity can be measured by the increase it can bring to the efficiency of the use of labour in that field. The advantage of treating the problem in this way is that it makes it possible to reduce it to a series of questions which are, to some extent, familiar in socio-economic or technological forecasting. These involve estimating the relative importance of different fields of activity to human needs, and the relative importance of different needs, which can be tackled using input-output analysis and similar techniques.

These methods have further practical advantages. Many research organisations throughout the world already function by using input-output analysis and similar analytical methods, and have therefore gained considerable experience in their use, as well as substantial statistical back-up. Some of this experience might usefully be applied to the project on research and human needs.

But these techniques, though they may make solution of the sub-problems easier, cannot by themselves provide a really integrated attack on the problem. For that, a new method will be needed. Mathematical modelling—even using a model of enormous complexity—cannot solve the problem either. A formal model is inappropriate when the question under analysis is characterized by many elements which cannot be quantified. But at the same time, a straightforward expert analysis will not suffice either; although such an analysis may be better than a formal model for handling qualitative aspects, it cannot deal with the systematic interaction of many different elements. In addition, a pure analysis of this sort finds it difficult to incorporate the many possible development scenarios.

I believe a solution is to be sought in the combination of expert and formal methods of analysis. One possibility might be to use a problem network analysis technique which has been developed over a number of years by the Institute for World Economy and International Relations of the USSR Academy of Sciences. This method, which uses both human and computer analytical methods, is designed to treat complex problems with many variables, qualitative as well as quantitative.

Human Needs and Global Development Models

N. I. LAPIN AND V. V. YURCHENKO*

One of the best methods for analysing possible development scenarios is the building of global models. The origin and growth of this field represents a new stage in man's understanding of the complex problems which will have to be solved if the needs of the great majority of the Earth's population are to be met.

There is a close link between global modelling and research into human needs, a link which shows itself in two ways. On the one hand, the analysis of human needs helps to establish the parameters for those engaged on building global models; while on the other global modelling, by examining possible development approaches in full knowledge of resource constraints, gives one an idea of what the long-term prospects are, and thus enables one to compare human needs more precisely and make better estimates of how likely they are to be satisfied.

Global modelling began by studying the possibilities of satisfying essentially physiological needs, such as food, a clean environment, health, and so on, and then moved on to an analysis of spiritual needs such as education. But this shift now appears to have come to a halt; in the most recent models produced in Western countries considerable attention is paid to the solution of discrete technological and economic problems without really considering the satisfaction of real human needs.

The needs of individual human beings, when considered in the widest possible sense, derive from social and cultural causes. Human needs acquire a really human nature only when seen in a social context. The problem is made more difficult to analyse by virtue of the fact that today many different societies exist side by side, creating for their peoples substantially different needs; and, at the same time, because of growing interdependence between nations, the degree to which these needs can be satisfied depends increasingly on the development of the whole of mankind.

Dissatisfaction with the present global models springs from the fact that they reduce the whole complex of human needs to a small set of mainly physiological ones. This dissatisfaction has shown itself in the efforts made by

* Dr. N. Lapin and Dr. V. Yurchenko are at the Institute for Systems Studies, USSR.

some people to analyse global problems by traditional humanitarian methods within a formal structure. By doing this, the creators of such models have been able to enlarge the scope of the analysis by including such needs as "a just social order" or even the need for universal revolution. But such needs have remained very dim and diffuse, as have the real social forces capable of bringing them about. In other words, the rejection of the formal methods in favour of more traditional approaches has not necessarily produced any more convincing results, or been accompanied by the discovery of any new ways of meeting human needs more effectively.

Global simulation models thus face the problem of including social questions more fully than did the first generation models. One problem that inevitably arises is the need to reduce criteria of quality, which are vector quantities, into scalar quantities for incorporation in the model. As a rule, such reductions are made formally, a process difficult to apply with social questions which are not easy to formalise. In our opinion the key to solving this problem lies in setting up realistic scenarios which allow us to display realistically world development alternatives for the next 20–25 years.

A worthwhile step in this direction has been made by a group at the Social Research Department at Sussex University headed by S. Cole.[1] This group has developed a system of scenarios of world development which take into account the significant socio-political forces of the present day. A drawback is that the system has been developed in relative isolation without reference to a global model which could include all relevant parameters.

In creating development scenarios, the first step in our view ought to be the creation of a structure of world development aims, expressed in very general terms and on the assumption that alternative aims also exist. This set of aims, altered and modified to reflect the interests of the principal socio-political forces, should form the basis of the development scenario. A simulation set up in this way to reflect formalized as well as non-formalized elements allows one to carry out experiments by searching for global solutions to problems meeting the needs of the greatest number of people.

Note

1. S. Cole *et al.*, Scenarios of World Development, *Futures*, February 1978.

*An Action Programme on Research and Human Needs**

B. M. UDGAONKAR†

In discussions about basic human needs, there is often a certain inclination to consider them from a philosophical point of view. There is always a tendency among intellectuals to try to sharpen definitions and to use more and more refined tools of analysis. While such exercises at the analytical and philosophical levels are certainly not useless it appears to me that the emphasis of the UNESCO programme should be on action leading to the fulfilment of those basic survival needs which are today denied to almost 30 per cent of the world's population.[1] Those who are below the subsistence or survival level are not interested in refinements of definitions. What they want is urgent action.

The key question is whether international action, and research directed towards such action, can save the 1000 million people below the poverty line who today are essentially written off by the national and international systems. Sensitive people have even claimed that the condition of these 1000 million people is no better than that in the concentration camps of World War II, and that the more fortunate ones who do nothing about it are as culpable as those who did nothing when millions of their countrymen were dying in those concentration camps.[2]

Understanding Human Needs

At a philosophical level, human needs are closely related to the values which underlie human existence, to the purpose or goal of a person's life. Maslow[3] defines this as "realizing the potentialities of the person, that is to say becoming fully human, everything that the person can become". Similar ideas have been expressed by ancient philosophers, in India and elsewhere.

There is, however, a wide gap which separates this perception, and the human needs derived from it, from the view of an individual and his perception, which is largely material in nature. Several authors[4] have drawn attention to the distinction between needs directed towards self-actualization

---

* Report prepared by the author, as a co-ordinator of a Working Group set up UNESCO, as a part of its Programme on Research and Human Needs.

† Professor B. M. Udgaonkar is at the Tata Institute of Fundamental Research, Bombay, India.

and those which are based on consumerism; between needs which lead from within man, from his striving towards self-realization or fulfilment, and the needs of the market, or those that the market manipulates. On the more practical plane, therefore, there are different approaches to the definition of a hierarchy of human needs.

There can, however, be no disagreement about some of the *necessary* conditions without which human fulfilment is impossible. First among these is the need for survival—for individuals, communities, nations and cultures. Related to this "survival line"[5] are, on the one hand, the individual's needs for food, water, clothing, shelter, sanitation and health services, education and the right to work or an income-producing occupation; and on the other hand, the collective need to preserve world peace and to create conditions in which nations and cultures can develop in an autonomous and self-reliant manner. If international action in this area, through UNESCO in particular, is to make a dent on the problem, it can only be if human solidarity becomes much stronger than it is today. In this context, one may list human solidarity as a basic need of prime importance, and any action-research programmes that promote it would be extremely valuable.

Efforts to understand and define human needs in the round are worthwhile because, first, they enable us not to lose sight of the higher (non-material or spiritual) needs while attending to the urgent problems of the survival needs; secondly, they may lead to action within the industrialized countries to depart from the present pattern of wasteful growth, and to evolve new life styles which are not only good for themselves but would also present an alternative model of development for the poorer countries; and thirdly, they lead to a fresh approach on alternative development styles among the poorer countries themselves, an approach which may help the richer countries in their search for alternatives.

For example, one may consider possible steps that might be taken to meet the basic need of food. A person's food needs may be met either through unemployment pay or charity of one kind or another, or on the basis of a social policy which guarantees the right to work. This latter would at the same time provide the individual with a sense of security and human dignity.

One must thus define priorities in needs, and in the programmes for their satisfaction, so that in satisfying one need, one is already laying the groundwork for the satisfaction of other needs somewhat lower in the priority list. Such an approach is characteristic of a holistic view of things.

One must not forget that human needs evolve. One of the principal human needs is activity—material, intellectual, social, political, creative and so on. These activities themselves give rise to further needs, and it is important to develop social and cultural indicators that take account of all these needs. They may help us define a dynamic approach for international action in this area.

It is also important to distinguish between needs and **pseudo-needs.** A large fraction of society's resources is today consumed by pseudo-needs. This question, though complex and sensitive, needs to be studied by behavioural scientists with a view to identifying possible steps that may be taken by society to rid itself of pseudo-needs in the long run.

The data base for meaningful research into human needs is inadequate, even as regards elementary material needs. Some data exists as a result of the work of agencies like WHO, FAO, ILO, UNDP, IBRD, etc., and several valuable sectoral investigations have been carried out by such agencies. But accurate and reliable information is still needed on such topics as: the distribution of availability of food, health care, water supplies, clothing, shelter, energy, and wages and incomes. Other areas of ignorance include the extent of unemployment and underemployment, of literacy and education generally, and of libraries, in rural and urban areas.

Reliable data on these and other indicators such as birth rates, growth-rates of children, child mortality, body-size of adults (height and weight), morbidity and life-expectancy, death rates and causes of death should make it possible to approach the difficult question of the definition of indicators and norms, and of alternative development strategies. At least some of these norms for human needs may differ from society to society, and will also change with time. Needs change because there is a feedback between the satisfaction of needs and their perception; in addition, steps taken to ensure the satisfaction of certain needs (for example, food) may set in motion various transformations in the society (modernization of agricultural practices, and land laws, linking of village with the larger society, and so on) which may in turn have effects on the perception of other needs (size of family, health care, housing, education) or create yet other needs (fertilizers). While it is difficult to forecast how human needs will develop in future, it is desirable for societies to recognize that they will change, and try to forecast the likely directions. UNESCO, through the Committee of Advisers, may assist in such efforts.

While this information is being collected, it would be useful for the Committee of Advisers to commission a critical survey of the existing literature on indicators and norms and to encourage work on the development of non-material indicators of social well-being and on tentative models of the structure and pattern of fulfilment of human needs. It would also be useful to commission an international **source book** on the problems of human needs, standardized terminology, and a bibliography of available literature. A **Newsletter** on on going research and action programmes in these areas would also be useful. In fact, UNESCO might first start a Newsletter using the material appearing in it as the basis of a source-book later.

World Problematique and Human Needs

The problem of basic human needs cannot be isolated from that of world peace. In a paper presented at a meeting of the European Group on Research and Human Needs, Danzin[7] has posed this problem very eloquently. Danzin uses the Pestel-Mesarovic model to project the demographic scenarios that may evolve in the course of the next 50 years, in the rich and the poor parts of the world; he points out that the proportion of world population in the rich countries may come down from 37 per cent in 1950 to 20 per cent in 2000 and possibly as low as 10 per cent by 2050; that in that case, there would be a growing demographic pressure on the southern frontiers of Europe and the United States; and he suggests that if the basic needs of the LDCs remain unsatisfied, the young in these countries may consider that they have nothing to lose by violence, and that they may organize various forms of violent protest which would be difficult to control. He then asks if the probability of such a development should not induce a feeling of solidarity of interest among the industrialized countries. Such a solidarity, he adds, could be understood in several ways, but only the humanistic point of view can be examined under the framework of UNESCO: how can we reduce tensions and bring hope of satisfying basic needs for the poor people of the South?

Analyses such as this bring home the realisation that we may be on the verge of one of the great discontinuities[8] in human history. They also indicate that the time for taking decisions is very short (and getting shorter). While it lives on borrowed time, society has yet to learn how to face and solve the growing number of problems it faces. Could the mass-media not be harnessed to increase the sensitivity of people and to force the pace of decisive action, without which the problems may get out of control?

The priorities for research and action have to be defined within this perspective. The prime criterion for the selection of a research and/or action programme should be the fastest possible satisfaction of the human survival needs in such a way that as to lead to the satisfaction of "living together", "achievement", and "perfection" aspirations as well. Another criterion is that the dependence of the affected population (internationally or within a country) must be reduced. Important indicators of the success of such a programme would be the extent to which life-expectancy at birth has been raised, child-mortality brought down, literacy raised, and dependency reduced.

It is generally argued that it is not possible for the LDCs to hope to attain the level of consumption of the most developed country (the US), and therefore they must evolve a low consumption style of development. Furthermore, even if LDCs are content to remain at their present levels, and even if the material resources of the world are not close to exhaustion, the developed countries will not be able to continue their exponential growth in consumption of material goods very long, so that they also have to change their wasteful style of development.

In any case, increases of material prosperity do not necessarily add to the quality of life. If all this is true, a new type of growth, growth in the quality of life rather than in mere material prosperity, has to be evolved. The developed countries have a responsibility to give a lead, since they set the fashion which the less developed countries, rightly or wrongly, try to emulate.

The LDCs will also have to try to find such alternative routes to development themselves, irrespective of what the developed countries do. But it must be recognized that if the rich countries continue along their present path of wasteful growth, the problem of discovering an alternative path becomes all the more difficult for the poor ones.

Evolution of a new pattern of growth demands a better understanding of human needs, especially the spiritual ones. In the case of the LDCs, it also demands the creation of a strong scientific and technological infrastructure, which alone can provide them with the technological and intellectual independence, and the capability and courage essential for embarking upon a new style of development on their own. A development dependent on aid and technology from the rich can only lead the LDCs to pointless emulation leading to intensified stresses and strains.

Since energy and material resources are declining, the industrialized countries should try to identify, in the words of Etzioni:

> "pursuits which we truly enjoy, are low in energy, capital and work demands, and to expand those pursuits. Cultural pursuits are one candidate; they do not abide by the laws of scarcity. Unlike a snowmobile, a sailboat, a steak or any other source of material pleasure, one person's enjoyment of folk-songs, poetry or prints does not reduce the availability of these pleasures to any other person. . . . The same may be said for trips to the *inner* space . . . the main energy they consume is psychic."[9]

To return to the problem of meeting basic needs of the 1000 million people below the survival-line, what is mainly needed is a strategy which utilizes existing resources and knowledge, rather than trying to discover new knowledge. Results of new research may help, but one must avoid the risk of becoming a victim of a technological fix.[10] In particular, initiation of scientific research on basic needs, or normative research relating to the definition and understanding of basic needs, must not provide alibis for inaction in other areas.

It is therefore important to draw a clear distinction between those problems which can be solved immediately with existing scientific and technical knowledge, but where we have not been able to muster political will, and those problems which need more research. Among the latter, we may list the following: improving the productivity of tropical soils; nitrogen-fixation through biological or chemical methods; biological control of pests; photosynthesis; alternative sources of energy, especially those suited for decentralized exploitation; better understanding of parasitic diseases. All research, except in areas such as these, must be action-oriented or coupled with participatory action.

The problem of meeting basic human needs is social, economic and political. It is related to the uneven distribution of power, both between countries and within a country; it is a problem of social justice; it is a problem of political will and of values in the society in general and among the intellectuals and in the science and technology community in particular. It is a problem of the inner limitations of man rather than the outer limitations of natural resources and technology. It is also a problem of introducing structural changes within a nation and in the international system. It is *not* a problem of science and technology per se. UNESCO would be performing a valuable service if it focuses attention on this aspect, so that the responsibility for action may be located where it belongs. If one focuses too much attention on the science and technology aspects or the hardware aspects of the problem, one may create the false impression that the problem is being solved and thereby provide alibis for inaction on the socio-economic-political front. One must not help the decision-makers to hide behind supposed technological difficulties when the real difficulties are in the social and political domain, or to take a technocratic approach which leaves the core of the problem untouched.

The exisiting constraints have to be understood in behavioural terms in order to understand the true nature of the problem and lead towards action. For example, take food. An IFIAS report[11] states: "There is already enough food produced today to give *everyone* of the 4000 million people on earth *more* than the survival needs. But we all know that one-third of the world's population is having more food than is healthy and another one-third is constantly starving. . . . This is a *problem of ethical nature, political will and institutional constraints*".

Again, at the IFIAS Workshop on "Interaction of Health, Nutrition and Education in Optimal Human Development" held at Geneva in May 1976, it was remarked[12] that there is no mystery about what should be done. The basic problem is *not* the effects of malnutrition and poor health but the elimination of those effects. The key question was motivation of *action* rather than motivation of further research.

Today food is produced not to meet the needs of people who are starving, but in order that it may be sold. It is not uncommon to find the practice of deliberately destroying food in order to maintain prices,[13] or of subsidizing farmers to produce less.[14] Also one finds the peculiar situation[15] that food is stored as surplus (or even exported because nobody knows how to store it) even in countries where a large fraction of the population is below subsistence level. In such countries it should be imperative to define grain requirements (or food targets) in relation to nutrition needs of the population (for example, 150 million tonnes for 600 million people in India), and not in relation to its purchasing capacity. It is cruel to adopt the latter criterion and talk of a surplus when millions are going hungry. The problem is obviously a problem of economic systems, of international politics, and even more strongly of national politics.

Of course, it would be foolish to claim that research on protein malnutrition in early childhood and its irreversible effect on mental development is of no interest. One has, however, to consider the problem in its proper perspective. For example, it has now been established that in India, traditional foods, if taken in adequate quantity to provide the necessary calories, are also capable of fulfilling protein requirements.[16] Protein malnutrition, therefore, is simply a consequence of too little food, and thus is a problem of poverty. In other words, the problem is very largely socio-political. What it calls for is urgent action to feed people, and not more research. (An exception might be made for research designed to reverse or compensate for the effects of early malnutrition.)

Good quality foods—fish, fruit, vegetables, even sugar—are today being exported in increasing quantities from the poor countries to the rich ones.[17] Action to prevent such exports could be useful for improving nutrition within the poor countries. A similar problem often exists within a country—for example, the flow of milk and vegetables from villages to cities. The problem is one of economics and politics and of distributive justice, not of scientific research in nutrition.

As for the effects of malnutrition, the oft-repeated statement that protein deficiency can cause mental retardation in the human infant actually appears to be unproven. In human beings, it has been suggested that enrichment of environment is of equal importance to increased protein intake.[18] With few exceptions, of course, the great majority of the populations concerned suffer from cultural and educational deprivations as well—"social malnutrition" it might be called. So any investigation intended to demonstrate the relation of malnutrition to development must take into consideration all possible factors and not get lost in a technological-medical fix. UNESCO might usefully promote holistic studies of this problem.[19]

Another example is health care. The fact that up to forty times more children die between the ages of 1 and 4 in poor countries than in rich, or that 40 per cent of all deaths in the LDCs occur among children under 5 years of age, is the result of malnutrition, poor sanitation, inadequate maternal care by ignorant mothers, and inadequate medical services. That is, it is a consequence of poverty and not of lack of knowledge. They are victims of diseases that modern medicine could perfectly well cure or even eradicate. More generally, a large fraction of deaths in the developing countries is due to communicable diseases and gastroenteric disorders, which are in turn related to inadequate water supplies and sewage disposal. These, too, are not the result of ignorance but of poverty. They are also a result of misplaced priorities. Spending in these countries ought to be directed to public health services designed to prevent diseases rather than to cure them, but preventive medicine tends to be treated as a second-rate field.

Disease and death in poor countries is therefore caused as often by socio-political and economic circumstances as by medical causes. For when the cause is medical, its solution is available but not applied, because of economic and social constraints. The affected persons are victims of society rather than of disease. Medical services are only one of the many factors which determine the health of a population. Health status is also determined by the availability of other social services, like water, drainage and environmental sanitation, education, especially of women, and employment without which nutrition remains poor.

Martin Kaplan[20] has drawn attention to the fact that "assistance from advanced nations has often adversely influenced health activities and the development of health services in the poor countries, especially where bilateral aid programmes are concerned. These programmes with their concern for visibility not infrequently include spectacular but ill-adapted show-pieces, such as large hospitals and advanced technical apparatus, and training fellowships in other countries in highly specialised aspects of clinical and surgery. . . . *Such actions tend to perpetuate the dependence of the poorer nations on others, rather than encouraging them to achieve self-sufficiency to the greatest extent possible through self-reliance—which is true independence*" [my italics].

Parasitic infections like malaria and schistosomiasis have been called the cinderella of biological problems, even though hundreds of millions of people are affected every year. The understanding of parasitic diseases, leading to the development of vaccines is a basic research area of vital importance, to which WHO has fortunately started focusing attention.

Malnutrition, disease and mortality in childhood are largely socio-economic and political problems. But even after the socio-economic and nutritional problems are taken care of, children will still be subject to gastro-enteric and infectious diseases unless simple hygienic measures are taken by the mother herself. Education of mothers is therefore important. This education has to be given in the appropriate local socio-cultural context. UNESCO may therefore support pilot action-research projects in the area of educational, cultural and religious aspects of the education of mothers in child-care, in relation to its impact on child mortality and morbidity on the one hand, and on birth rate on the other.

Another example comes from the domain of drugs. Again and again it has been pointed out that the mass production of about 100 or 150 commonly-used basic drugs and their sale under generic names rather than brand names would go a long way towards making drugs available to the needy poor, at prices which they can afford.[22] The action here again lies mostly not in the medical domain but in the social-political domain, and is complicated by the existence of strong trade interests, in particular those of multinational corporations.

One more example may be taken from the world of education. The eradication of large-scale illiteracy is a problem of mobilising available resources including human resources, more than anything else. Gunnar Myrdal has remarked: "When a country becomes communist, a vigorous campaign is usually waged to make the whole people literate within a few years." However, as he goes on to add, "There should be nothing sinisterly communistic about this particular policy".[23]

Returning to agriculture, an important factor is the problem of land tenure and ownership and the related problem of the availability of technical services and cheap credit to small farmers. New high-yielding varieties accompanied by appropriate inputs of agro-chemicals have a value, but their limitations are seen from the uneven production of grains in countries which have adopted these varieties, and also the fact that yields on real farms do not approach those gained under controlled conditions on experimental farms.[24]

Processes of technical and social change in a community are interdependent, and efforts in both directions have to be simultaneous and co-ordinated. Attempts to introduce techniques alone, without a proper perception of the local scene, are often unsuccessful. The uneven results of the high yielding varieties of different cereals are an example.

The poverty of developing countries is often blamed on the population explosion. But it is difficult for a rational population policy to work in a situation where child mortality is high, literacy is low and social security non-existent. On the contrary, the Latin American World Model[21] suggest that the only adequate way of controlling population growth is by improving basic living conditions. And an ILO report[25] emphasizes: "High birth rates in poverty-stricken areas are not the cause of underdevelopment but a result of it". What is needed in such a situation is an urgent effort to make health services available, to bring down child mortality and to improve the quantity and quality of educational opportunity in the rural areas, and among the lower income groups in the cities, with special attention to women. The influence of such factors on birth-rates is seen from the example of Kerala in India.[26]

The UNESCO decision to launch a programme on Research and Human Needs was taken against the background of the search for a New International Economic Order. Through its holistic approach, the UNESCO programme can be distinguished from other programmes are carried out by sectoral agencies like WHO, FAO, ILO and UNDP. In other words, the UNESCO programme should focus itself on an *understanding of the causes rather than the symptoms of a sick system,* in which basic human needs remain unsatisfied for a large fraction of humanity. In doing so, we must remember that time is running out, and that there exists, as Feld[28] has remarked "an almost universal feeling of disillusionment with past efforts among the developing country scientists". This implies, in particular, that important research areas should deal with the causes of world poverty, especially the international and institutional causes, *and fundamental institutional reforms, mainly at the international but also at the national level,* that are needed to remove these causes. This indeed is the basic aspect of the proposed New International Economic Order.[27]

In this context one may note that the proponents of a New International Economic Order suggest that we are dealing with *a crisis of international structures,* wherein even resources which legitimately belong to the countries of the developing world are actually not accessible to them. They claim, for example, that the institutional disorders of the present world order imply a "loss" of economic opportunities to the Third World which may well be over 100 billion dollars a year.[29] If this is true, then merely by changing the institutional structures, so as to provide for increasing equality of opportunities for the poorer countries,[30] enough resources should be generated within the poor world to finance its development. But changes of institutional structures are not easy to bring about, and a worthwhile area for study and research under the auspices of UNESCO, might be how to bring about these changes by mutual agreement between the developing and the developed countries—and to do so fast enough, since we are living on borrowed time.

There are obvious difficulties in transferring to LDCs existing resources or even processes which are being commercially exploited in the industrialized countries. It is, therefore, sensible to discuss the possibilities of generating resources for development or for meeting basic human needs in other ways. For example, one might ask whether potentially rich but so far largely unexploited resources are being earmarked for development. Possibilities are sea-bed resources (which have been described as the common heritage of man) and the rich possibilities opened-up by recent developments in genetic engineering through techniques of recombinant DNA. In today's world, the tendency is for such developments to be made by the profit-motivated organizations, in particular by multinational corporations in the industrially advanced countries. It might be helpful if new international institutional structures, such as an International Authority for the exploitation of the sea-bed, and another for the exploitation of the advances in genetic engineering, could be set up so that such resources may become available for furthering the programme of basic needs. More generally, could UNESCO help to evolve a code of conduct for scientists, under which they will refuse to surrender their intellectual products for exploitation by profit-motivated transnationals? They could surrender them instead to an international authority, in which such intellectual property should vest.

The problem of relative poverty and destitution also exists within countries. In India, for example, the Planning Commission estimated some years ago that if the bottom 30 per cent of the population were to achieve a consumption level above the poverty line (defined in 1973–4 as consumption per capita of only $60 per year) it would require the top 70 per cent of the population to cut their consumption by about 10 per cent. This "sacrifice" cannot be considered excessive. Making the same comparison internationally, the sacrifice required from the richer countries is even smaller. For example, if the industrialized countries had kept aside only 2 per cent of their *increment* of income during the seventies (estimated at $1000 per head) for the benefit of the LDCs, the United Nations target of 0.7 per cent of GNP for development of LDCs would be reached. But, as we know, such sacrifices are not easy to bring about. A useful piece of action research might be to ask what type of education or other societal processes—if any—could help create the background for distributive justice. How could the media be used for this purpose? Alternatively, one might ask whether it is in fact possible to achieve distributive justice and to satisfy the basic human needs of the vast majority of people without class conflict within a country and confrontation if not wars internationally.

Disillusionment, arising from the widening gap between performance and promise, has undermined the faith of people in the democratic process, and in the effectiveness of international co-operation. The abolition of poverty and the establishment of social justice, even socialism, have now become the slogans of almost every political party and government round the world. Internationally we have had the Development Decades, the World Plan of Action, and various other marvellous programmes and plans prepared by "experts", aimed at the improvement of the human condition, which have fallen far short of their high expectations. The UNESCO programme on human needs should analyse the weaknesses of these past efforts and be so organized as to avoid these pitfalls.

Why is there such a big gap between promise and performance? Is this to be treated as an uncomfortable question to which we are incapable of finding an answer? Can one change the situation and bring about greater human solidarity through education or through the continuous sensitization of the intellectual community? How? By insisting on participatory studies and research and action among the poor on the part of every academic? *Is that to be the new role for the intellectual,* especially in the LDCs? Will they be able to play it? Is their alienation standing in the way? If so, how to de-alienate them? A host of such questions arise; and it will therefore be useful for UNESCO to encourage and support research programmes in universities[31] which involve some participatory action by students and teachers among the poor and needy.

UNESCO may also encourage and support programmes which involve universities and research institutions in the problems of integrated rural development, with a district or block as a unit. Development programmes have usually left the poorer and weaker sections untouched. UNESCO-sponsored action-research programmes may therefore be focused on the human dimension of development. Within every such programme there must be a built-in component to assess its effectiveness from the point of view of the common man who is the supposed beneficiary. Social scientists should therefore be associated with any such programme from the beginning, to monitor and assess its impact on the lowest 30 per cent of the population involved.[32] Such programmes will obviously have to be carried out separately in different countries and regions. But care must be taken to ensure that they are so conceived that they may provide replicable models and not remain mere show-pieces.

During the past few decades, science and technology have provided humanity with tools which, properly applied, could not only help meet the basic human needs of food, clothing, housing, health and education of *all* people on this globe, but also improve the material as well as the spiritual quality of their life. But instead, vast scientific efforts and technological resources continue to be diverted towards destructive applications. The magnitude of weapons research is often not realized: it involves almost 50 per cent of the science and technology manpower in the world and an expenditure of over $330 billion per year, which is twice the amount the world spends on education, three times what it spends on health, and about thirty times as much as the rich countries provide as aid to the poorer countries.[33] Meanwhile starvation, misery and destitution remain the norm for a large part of the world population. The world's resources are being squandered for the accumulation of greater and greater lethality, with little regard for present inequities or future needs. Could the resources not be redeployed to enable all men to live as men?

Development, or the satisfaction of human needs, is a challenge of a magnitude comparable to that of a major military campaign. Properly organized development reduces international tension and could therefore be regarded as a security investment. Against this background, and considering also the operational capability and the managerial competence of the military (including scientists and technologists engaged in weapons programmes) the forces represent a resource for development. In fact, a systematic redeployment effort could have a dramatic effect. How is such redeployment to be structured? Can one think of a gradually increasing armaments tax? The world's military expenditure today exceeds $300 billion per year, and even a small tax of 5 per cent would yield $15 billion per year for world development. Can one also think of a phased redeployment of the scientific manpower which is wasting its scarce brain-power on armaments? Today, the scientific community itself has a vested interest in the armaments race, since about 50 per cent of it depends on it for their livelihood. Could one devise a phased programme to reduce this from 50 to 45 and then to 40 to 30, to 25 per cent by the end of the century? Could one also develop scenarios where the world military expenditure is reduced in phases from over $300 billion to almost 0 by year 2000, with a suitable redeployment of men, materials and resources? Such social inventions may seem unrealistic today, and most countries, including those that can ill afford it, will probably continue to waste scarce resources on military hardware. This has many causes, among them fear of the economic effects of disarmament. But this would not carry much weight if a systematic redeployment of the military R & D potential is contemplated. Redeployment schemes must therefore show convincingly that they provide a real alternative to the military-industrial complex. Trade unions must also be involved in such investigations. Redeployment is thus an area which should be subjected to careful international study under the auspices of UNESCO.

In this context, one may also mention the Third World suffers from the backlash effects of the armaments race of which the prime movers are the industrialized powers. Even if one ignores the millions of deaths in a hundred odd wars fought during the last 25 years on the soils of the developing countries, with weapons largely supplied by the industrialized countries, the opportunity costs in education, health and economic development felt by the LDCs have been very considerable. Could the LDCs extricate themselves from entanglement in these backlash effects? This would also have a bearing on their capability to generate resources for an attack on the problem of basic needs.

Satisfaction of basic human needs within a country is a part of the development process, and the developing countries will, by and large, have to carry out this task *on their own.* They cannot expect substantial support from the developed countries at least until the they have solved their own problems of stagflation. In any case, as the Latin American Model[71] shows, aid will make a difference of only a few years. The LDCs cannot expect the outside world to assume their burdens, and will have to evolve strategies based on self-reliance,[6] including the collective self-reliance of the Third World countries among themselves, based upon mutual assistance and sharing of experience.[34] They will have to carry out internal structural reforms which may be necessary in order to ensure equitable income distribution and social justice, without which the basic needs of a substantial fraction of the population may remain unfulfilled, in spite of economic growth. In fact, pressure for reforms in the international order will have to be accompanied by corresponding reforms in the national orders, not only to derive maximum benefits from the proposed New International Economic Order, but even to develop the strength necessary for achieving it.

For a nation, the right to develop in one's own way in a manner consistent with one's own cultural heritage is a basic need. How is this to be done in the face of economic, political and technological pressures in an unequal world? Self-reliance has not only material aspects related to technology and finance, but also deeper spiritual aspects related to the world-view of the society concerned and the path of development it may choose to follow.[35] But over the years development has come to seem a linear process along the same path of material growth followed by the developed countries. National priorities and development plans often get distorted by aid, political pressures and the cultural invasion that occurs through the mass media. Aid and assistance can either help self-reliant and autonomous development of perpetuate dependence and underdevelopment, and it has been said that 70 per cent of international aid is of the latter kind. Case-studies of good and bad examples of international assistance from the point of helping self-reliant development, would be extremely useful for defining policies.[36] By promoting such case studies, UNESCO will be able to help the fulfilment of basic needs not only at the individual level but at the societal level of self-reliance. This is all the more necessary since the rich countries today are not, by and large, taking a sympathetic attitude towards the self-reliant stand of the LDCs, and often tend to dismiss self-reliance as mere rhetoric.

The path of self-reliance is not easy, especially for a country which adopts an open-door policy. No development is possible unless the intertwining relationships among the complex of problems at the global, regional, national and local level are analysed within a holistic approach. In addition, the socio-economical and political institutional structures at each of the four levels must be analysed, and processes evolved by which these structures may be transformed and made more effective in meeting human needs. Then and only then could we hope to build a world community composed of nations meeting human needs in a self-reliant way.[38]

The importance of self-reliance in the context of human needs has been emphasized by several authors.[37] Self-reliance means counting on one's own forces and potentialities. It is the opposite of dependency: it implies relations of parity and equality, or reciprocity between individuals or countries. It contributes, therefore, to what Albert Tévoédjrè calls the "contract of solidarity",[39] a solidarity which goes both ways and not just in one direction.

In this sense, *self-reliance and international co-operation are mutually complementary*. True "interdependence" demands self-reliance.

In the present international climate, however, interdependence appears as a hollow slogan to most people in the LDCs and the credibility of international scientific and economic co-operation is increasingly at stake. There have been serious suggestions (for example the COCOYOC declaration,[40] or Dar-es-Salam Pugwash Symposium on The Role of Self-reliance in Alternative Strategies for Development)[6] that conditions may arise in which a developing country may be forced to detach itself temporarily from the international scientific, educational, technological and economic system, so that it may develop according to its own genius, returning later to the international system on its own terms. Questions relating to improvement of the climate for international co-operation in science and technology, and other areas, including possible guidelines,[34] would constitute a useful area for action research. Such guidelines are being investigated under the auspices of Pugwash.

While one may hope that international action on the problem of basic needs will increase in coming years, the problems in today's world have to be solved on a national basis. Indigenous national capabilities have therefore to be built up rapidly. A research area of considerable interest would therefore be how to develop R & D infrastructure and scientific and technological capability in small developing countries in such a way that they may be in a position to be self-reliant in at least some areas, to plan their development autonomously, to make the best use of their natural resources, and also to contribute effectively to the collective self-reliance of the developing world.

A related problem is that of training scientists and engineers in a manner consistent with a country's priorities. Foreign training, or even training in the home country but following a pattern copied from the industrialized countries, often creates a problem by educating an élite better suited to function in one of the developed countries than in their own. A considerable effort is often needed to wean them from their intellectual dependence on the developed world. In view of this problem, UNESCO may emphasize training programmes which utilize training facilities in other LDCs rather than in DCs. Also, in building new universities and research institutes, LDCs may draw upon the experience (including mistakes) made by other LDCs.

Many problems relating to basic needs involve more than one discipline. But if a country is to be able to undertake high-quality transdisciplinary research, or be in a position to make a correct choice of technologies or development strategy, it must also possess sound disciplinary research in appropriate areas. Not much work has been done on how to create such a science and technology infrastructure, especially in small countries. UNESCO should therefore support some studies to work out possible models and strategies.

In this context a strong case can be made for supporting basic research in the developing countries. It is important to emphasize this, since LDCs are often criticized for carrying out basic research; in fact, encouragement of basic research is all the more important in the LDCs because they have such a slender base. High-quality basic research creates the capability for acquiring new knowledge on one's own, and fulfils the basic need of the community to be self-confident and intellectually self-reliant, and to develop a scientific culture of its own. In a society based on science and technology, no community can afford to derive its scientific culture from others. Moreover, when areas and problems are chosen judiciously, such basic research also builds up the capability for applied research. With this in view, we would like to recommend that UNESCO help the emergence of centres of excellence in the developing countries and in particular take an initiative in building up at least one International Centre of Excellence in a suitably chosen area of basic sciences, located in the Third World. Such a centre could play the same role for the LDCs as CERN has played for Europe, or the Tata Institute of Fundamental Research for India.[41].

A possible subject for such a centre of excellence would be radio astronomy. The centre could be built around a giant equatorial radio telescope, which could not only be used for fundamental research in astrophysics, cosmology, space physics and extra-terrestrial phenomena, but will also help to build expertise in the important areas of sophisticated technology such as microwaves and antenna systems, satellite communication and so on, as happened in India after the construction of an indigenously designed and fabricated radio-telescope.[42] The know-how exists in the LDCs and it could become a good example of collective self-reliance.

Several countries in Asia and Africa now have a reasonable number of scientists in different disciplines working in a large number of universities. Modern research demands research groups of a certain minimum size and viable groups therefore must be created in at least a few centres in these countries, if their scientists are to be enabled to make an impact on competitive international research. For example, there are about 200 mathematicians with doctorates in Africa, but they are handicapped by distances and a sense of isolation. UNESCO may help create a network of communications among these scientists, through visiting fellowships and possibilities of spending some months a year in a few suitably strengthened centres of excellence within the developing countries, thereby enabling them to function at their best.

Development is not only of people but by people, with priorites set by them, so literacy programmes for the 50 per cent of the population who today are illiterate must have a high priority. UNESCO might usefully promote an analysis of why adult literacy programmes and programmes of universalization of primary education have failed in so many countries so that a more effective policy for eradication of illiteracy may emerge.

Literacy is only part of the educational problem. In many of these countries there is also a problem of language—the fact that the language in which science is taught and practised is foreign to the vast majority of the people. This creates an obvious obstacle to the spread of a scientific temper: it makes science appear as a benevolent but mysterious activity of the élite for the "uplift" of the poor.[43] UNESCO might therefore promote the emergence of scientific literature in the languages of the ordinary people.

But to achieve participation of large populations in planning, more than education is needed.

The ILO report mentioned earlier recommends:

> "The main thrust of a basic needs strategy must be to ensure that there is effective mass participation of the rural population in the political process in order to safeguard their interests. In view of the highly hierarchical, social and economic structure of agrarian societies in some developing countries, measures of redistributive justice are likely to be thwarted unless backed by organization of rural workers. A policy of active encouragement to small farmers and rural workers' organizations should be pursued to enable them to participate effectively in the implementation of: (a) programmes of agrarian reforms, distribution of surplus lands and land settlement; (b) programmes for developing ancillary services such as credit, supply of inputs and marketing; and (c) programmes concerning other employment generation schemes, such as public works, agro-industries and rural crafts."[44]

Such a strategy is bound to involve radical structural changes in traditional hierarchical societies and is therefore likely to be resisted in most countries. What is involved is a task of social systems engineering. It would be worthwhile for UNESCO to promote some pilot action-research programmes in this area.

Employment is another area deserving special study. How can we evolve an employment and production strategy in which the large fractions of the population who are today unemployed or underemployed may find opportunities for work, so that at least their basic survival needs of food and shelter can be met? In other words, how can we establish, guarantee and implement the basic human right of work? While there has been much discussion of appropriate or employment-intensive technologies (and of integrated rural development) no clear-cut strategy has yet emerged. This is, therefore, an area in which action-research should be encouraged.[45] The precise solution would of course depend on the size of the country, its stage of development and what social-political system it wants to follow. There are, however, several questions common to each situation. For example, what is the best mix of capital-intensive and labour-intensive technologies which will ensure full employment within a country, and how to achieve it? Does it imply that certain kinds of production be entirely reserved for the decentralized small-scale employment-intensive sector, and insulated from competition by the large-scale sector, ignoring the usual arguments of "economy of scale" (which ignore social costs)?[46] If so, how does one ensure quality control in such a form of mass-production, by the masses for the masses? Does such an employment-intensive strategy imply that within a country one must aim at self-sufficiency for regions or areas (say with 100,000 population, or 100 villages), if not for individual villages as envisaged by Gandhi? Gandhi dreamed of self-sufficient villages and Mao organized self-supporting communes. Is this possible in a free-market economy, with open competition from large-scale urban industries? What kind of social commitments does such a strategy imply?

Modelling exercises and pilot projects to define a "bootstrap" process by which the economy of a country could be lifted through such a mix of technologies would be very useful. They will help define alternative development strategies. In the same context, studies of energy needs of rural–urban societies, with the aim of meeting a substantial fraction of those needs in a decentralized fashion through solar energy, photosynthesis, bio-gas plants, wind-power and so on, would also be very useful. Pilot projects in this area could therefore be supported by UNESCO.

Human needs have been increasing for centuries. What was a luxury some centuries or even a few decades ago is often considered a necessity now—central heating or air-conditioning, motor cars, radio, TV, various domestic gadgets, and so on. In this situation, it is legitimate to ask whether one can define basic or minimum material needs at all when the sky is the limit. What are indeed the "real" needs of man? To understand the dimensions of the problem, one may note that the poverty-line in the USA is drawn at an income level which roughly corresponds, at official exchange rates, to the highest range of salaries in India.

A distinguished visitor from the Western world to India wrote recently about what he calls "a nightmare vision of the future".

> "In say 50 years time [he wrote] between a billion and billion and half of Indians with a GNP per capita allowing them to a fair degree to equip themselves in the way we have done over the last 30 years, which would mean over 100 million automobiles, one-sixth of the country covered by concrete and glaring transistor radios carried around by a billion people. *This is so evidently undesirable that the quesion of an alternative way of things is mandatory.*"

People in the poorer countries fail to understand why such a vision is described as a nightmare and will continue to do so unless they see evidence that the rich industrialized world is curbing its own consumption and "growth". Otherwise it appears to them that the advice they are given—to adopt new low-consumption living styles and corresponding development strategies—is only designed for the poor, and that poor countries are being asked to conserve their resources so they can be frittered away by the rich.

What are the steps (including lower energy consumption and less profligate living styles) that should be taken by the rich countries in order that the poor countries may be encouraged to adopt alternative, decentralized, low consumption living styles? Consumption maxima must be established along with minima, and the role of promotion agencies will have to be altered (no more promotion of consumption)! Action research in these directions should be encouraged in the developed countries. But mere considerations of solidarity will not take us very far. One must be able to demonstrate that consumption can be reduced without loss of style or comfort of living and without creating unemployment. Otherwise, the world will not move away from the present ruinous fashion of consumption, short of a catastrophe.

Several thinkers in the industrialized countries have been deeply concerned about the path taken by their own societies, and have emphasized the importance of discouraging the craze of people for the satisfaction of material needs or pseudo-needs, and of encouraging the pursuit of non-material or spiritual needs.[47] It should be possible to improve the quality of life, without being trapped helplessly in the growth spiral. On the other hand, this concern has yet to be converted into action. One cannot underestimate the difficulties, when short-term profit motives dominate the thinking of industrialists and their promoters in the mass-media industry; and when everyone concerned worries, not only about a loss in income or style and comfort of living, but also about his job security if growth were to be curbed. Problems relating to changes in present wasteful life-styles would therefore be a key area for action-research within the industrialized countries.

As Forti[48] remarked at the beginning of the background paper for the Venice meeting, "There is no doubt that humanity is at a turning point and that the traditional order of 'values' must go through a thorough revision", and also that scientists must contribute to "a radical analysis of our present-day problems". Scientists are said to be that part of humanity which is in league with the future. If so, the most important contribution they could make to the problem of human needs is the propagation of the new value system, creating a new pattern of international relationships and involving a transition from the colonial or neo-colonial relationships to a reciprocal relationship of a community of self-reliant nations. In this context, one may note the stresses to which the UN organizations, including UNESCO, have been subjected in recent years. What is at stake is human solidarity, and probably the best way to ensure and reiterate the utility of UN organizations is to launch a massive programme on human needs.

Underlying these discussions is the basic need for human solidarity both within a nation and internationally, and for a change of attitude among the more fortunate members of the human family (especially the intellectual élite) who are today largely alienated from the problems that matter. The re-awakening of the intellectual to values based on human solidarity is one of the transcendental needs.

An important question that needs to be studied by behavioural scientists, therefore, is that of motivational or value factors. While it has been realized for some decades now that science and technology have placed in the hands of man tools by which poverty, want and destitution could be eliminated, social action has lagged far behind. And while scientists agree at an intellectual level that research and action relating to basic human needs are important, in practice efforts in these areas do not enjoy the same prestige as fundamental research or even military research. The career structure and the approbation and reward system of discipline-based science are an obstacle to trans-disciplinary research and action and must be changed. It is tragic that special efforts are needed to persuade individuals and societies to act on problems that matter for humanity. Human motivational factors and problems relating to changes in the value system need to be studied much more than they are today, so that in the long run the problems that matter may get the attention they deserve. Such studies, unfortunately, are not fashionable; they are fragmented, and have hardly any impact on action at the social or political level.

In the meantime, in order to focus attention on the importance of research relating to human needs, to provide an incentive for such research, and to help shift research priorities from glamorous fields to fields that matter, UNESCO should allocate a substantial and increasing fraction of its funds to the promotion and support of research in this human needs area, and recommend member states to adopt a similar policy for national research.

Further, since the solutions of the problems often lie in the socio-political domain, it is necessary to develop ways of using the powerful mass media for sensitizing people to the urgency of the problems.

Is the science and technology community committed and ready to guarantee the fulfilment of basic human needs as a universal human right? Will it be able to translate this commitment into working for the things that really matter: the creation of an international food reserve; the denunciation of the use of food as a weapon, and of the life-boat ethic; the New International Economic Order, including a revision of the terms of trade in basic commodities; a Code of Conduct for Technology Transfer; the institution of programmes for changing consumption patterns in the rich countries and ensuring guaranteed minimum living standards for everyone in the poor countries; progressive cuts in the armaments expenditure in all countries and a diversion of resources and scientific efforts into the constructive direction of development; and the creation of a self-reliant science and technology infrastructure in and for the Third World?

Some Areas for UNESCO Support, through the Committee of Advisers

We will conclude by summarizing some important areas where basic research, action research, or modelling studies, under the auspices of UNESCO, will help the global involvement in the problem of basic human needs.

The *UNESCO programme should be centred around the focal point of the New International Economic Order, and attempt basic holistic understanding of the problem, going to the root causes rather than merely treating the symptoms of the sick system* in which basic human needs remain unsatisfied for a large fraction of humanity, even though the science and technology tools and resources available to man are capable of fulfilling the basic needs of the entire world population in spite of its recent explosion. This focal point must not be lost sight of, no matter what the research-action programmes be. The *major problems lie in the affective domain rather than in the science and technology domain per se,* and involve a complex interplay of science and technology and socio-political and behavioural sciences.

In our discussion and the following summary of areas for action-research, we have suggested some modelling studies. It must, however, be emphasized that results of modelling exercises cannot be better than the inputs or the value premises of those who formulate the models. They may be mathematically neat, with tantalizing computer print-outs, but may not conform to reality because of hidden or implicit value assumptions, or inadequate emphasis on the human dimension. It can also happen that the social objectives defined in the preamble to a model may not be adequately represented in the actual mathematical model which follows, thus misleading not only the common man but also an unsophisticated decision maker. It is therefore important

(a) that assumptions/inputs in a model be transparent for the benefit of the user;

(b) that Third World countries do their own modelling exercises, with software developed by themselves.

Another point that may be emphasized in relation to the UNESCO programme is that there cannot be a common world-wide programme on research and human needs, since specific action would depend on the local social, cultural, economic, political and technological situation. Above all, needs are as perceived and felt by the people themselves. The list given below may, however, help UNESCO as well as individual member countries to indentify action-research areas for support.

(i) Causes of world poverty and inequality, especially the international and institutional causes; and correspondingly, the fundamental institutional reforms, mainly at the international but also at the national level, that are needed to remove these causes.

(ii) The question of changes in international institutional structures, so as to provide increasing equality of opportunities for the poorer countries: what are the changes needed, and how to bring them about by mutual agreement between the developed and developing countries? Possible modelling exercises around this theme.

(iii) Research on lack of action: Why are the marvellous plans prepared by experts, nationally or internationally, not leading to action? In particular, an analysis of the weaknesses of the World Plan of Action from the point of view of implementation.

(iv) What types of education or other societal processes could help distributive justice? Is it at all possible to achieve distributive justice and satisfaction of basic human needs by the vast majority of people, without class conflict within a country and confrontation if not wars internationally? Possible modelling exercises might help.

(v) The question of evolving an employment and production strategy which would enable a country to establish, guarantee and implement the basic human right of work or income-producing occupation. Action-research in this area will be useful.

(vi) Pilot programmes for action research for development from below.

(vii) Participatory research programmes, on a pilot project basis, at some universities.

(viii) Modelling exercises to define a bootstrap process by which the economy of a country may be lifted through an appropriate mix of capital-intensive and labour-intensive technologies.

(ix) Study of energy needs or rural or rural-urban societies, from the point of view of the possibility of meeting a substantial fraction of them in a decentralized fashion through solar energy, bio-gas plants, wind-power, etc.

(x) What are the series of steps to be taken by DCs (including maxima on consumption) in order that LDCs may be encouraged to adopt alternative, decentralized, low-consumption living styles?

(xi) Role of mass media and promotion agencies in achieving such changes in living styles. Also, the role of education and mass media in sensitizing people to the urgency of the problem of basic human needs.

(xii) The question of promoting self-chosen and self-reliant development in the face of economic, political and technological pressures in an unequal world.

(xiii) Guidelines for international co-operation in science and technology for development, including programmes sponsored and procedures followed by various UN agencies.

(xiv) Case studies of good and bad examples of international assistance/aid/collaboration from the point of view of reducing dependence and helping self-reliant growth and development.

(xv) Question relating to development of S & T capability in a small LDC.

(xvi) Creation of network of communications among scientists in the LDCs, so as to remove their sense of isolation and promote the emergence of a few viable, critical-size groups.

(xvii) The creation of one or more Centres of Excellence in suitably chosen area(s), located in the LDCs, under the auspices of UNESCO.

(xviii) Human solidarity and the question of redeployment of a substantial fraction of resources (man and materials) from weapons programmes to programmes related to basic human needs; questions related to the revision of the value system underlying the organization of present societies.

(xix) Pilot projects on educational, cultural and religious aspects of education of mothers, in child-care in particular, in relation to its impact on child-mortality, morbidity and fertility.

(xx) Holistic research on effects of protein malnutrition and "social malnutrition" in early childhood on mental development of children, especially from the point of view of reversing or compensating for the effects by enrichment of environment or otherwise.

(xxi) Basic research on parasitic diseases, which may help develop new vaccines against scourges like malaria.

(xxii) Research in areas such as improvement of the productivity of tropical soils, nitrogen fixation, biological control of pests, photosynthesis, alternative sources of energy, especially those suited for decentralized exploitation.

(xxiii) Creation of holistic data base around basic human needs as a focal point; publication of a UNESCO Newsletter and Source Book.

(xxiv) Development of social and cultural indicators, for the degree of satisfaction of basic needs.

(xxv) Structure and inter-relations of human needs, aspirations and satisfiers; analysis of pseudo-needs from the point of view of long-term social action to get rid of them.

(xxvi) Steps to shift research priorities from glamorous fields to the problems related to basic human needs; changes in the research funding policy of UNESCO.

(xxvii) Promotion of new international institutional structures, such as an international authority for the exploitation of the sea-bed, so that resources may be available for furthering the programme of basic needs.

Notes

1. *Moving Towards Change: Some Thoughts on the New International Economic Order,* UNESCO, Paris, 1976.
2. Opening paragraph of the Report of the Meeting of the Advisory Group on Research and Human Needs, Moscow, 12–14 July 1976.
3. Abraham Maslow, *Psychological Data and Value Theory in New Knowledge in Human Values* (Harper Row, New York, 1959).
4. For example, Eleonora Masini, "Why Ethical Aspects in Future Studies?" WFSF preprint, June 1977.
5. We are here using the expression "survival line" instead of the expression "poverty line" used commonly in some countries; the line is to be drawn at an income level below which people cannot but starve. There are various estimates of the number of people below this line, but none seems to be smaller than 1000 million people.
6. See Appendix I for a discussion of the concept of self-reliance in relation to strategies for development. See also, *Pugwash on Self-Reliance,* edited by A. Parthasarathi, B. T. Feld, W. W. Chaǧula and P. J. Lavakare (Ankur Publishing House, New Delhi, 1977).
7. André Danzin, "Three Propositions concerning the LDCs and Four Questions-propositions concerning the Industrialized Countries", Venice Colloquium on Research and Human Needs, 25–23 April 1977.
8. Lester R. Brown, "The Discontinuities before Us", *The Futurist,* June 1975.
9. Amitai Etzioni, "Creative Response to our Crisis", *Bulletin of Atomic Scientists,* **33** (2), 24, 1977.
10. By technological fix is implied a predilection to seek solutions to social or socio-technological problems, including those of other cultures, in purely technological terms, ignoring or underestimating the human, social, cultural and political factors in which such problems are rooted. Ignoring software aspects can be a part of a technological fix.
11. *IFIAS Report,* May 1977 (stress added).
12. Report of the IFIAS Workshop on "Interaction of Health, Nutrition and Optimal Human Development", WHO, Geneva, May 1976.
13. See for example, R. Fieschi and U. Emiliani, in *Physics in Industry,* edited by E. O'Mongain and C. P. O'Toole, p.239, 1973.
14. Recently newspapers reported that the world grain production during 1976 was estimated to be about 1.3 billion tonnes, and that since this implied a "surplus" of about 200 million tonnes, USA was again being forced to subsidize farmers for producing loss.
15. A recent report on the Asian Agricultural Survey, prepared under the auspicies of the Asian Development Bank, points out that while good harvests of the past three years in Asia have filled up reserve granaries throughout the Far East, "in these same countries literally millions of people are still going to bed hungry in the same way they have been doing for as long as they can remember. They simply are unable to find an occupation which will bring in enough income to feed themselves or their families."
16. For example, C. Gopalan and B. S. Narasinga Rao, "Nutritional Constraints on Growth and Development in Current Indian Dietaries", *Indian Journal of Medical Research,* **59** (suppl.), 143, 1971. C. Gopalan *et al, American Journal of Clinical Nutrition,* **26,** 563, 1973. S. G. Srikantia, *Proceedings of the Nutritional Society of India,* **15,** 6, 1973. P. V. Sukhatone, *Indian Journal of Agricultural Economics,* **25,** 57, 1972.
17. For example during 1976-7, India exported fish worth more than $200 million to the rich countries, and these exports are increasing rapidly. The rich countries welcome the imports of food from the poor malnourished countries, but have been putting restrictions on import of finished goods like garments.
18. See, for example, Donald B. Cheek, "Nutritional Restriction and Brain Growth".
19. Cipriano A. Canosa, "Early Nutrition and Mental Performance", in *Pre- and Postnatal Development of the Human Brain, Mod. Prob. Paediat,* **13,** 256, 1974. The author wishes to thank F. Mayor Zaragoza for drawing his attention to the papers of D. B. Cheek and C. A. Canosa.
20. Martin M. Kaplan, *Bulletin of Atomic Scientists* **32** (3), 30, 1976.

21. *Catastrophe or New Society? A Latin American World Model,* Amilcar O. Herrera *et al,* IDRC—064e, International Development Research Centre, Ottawa, 1976.

22. See, for example, B. B. Gaitonde, *Seminar* **190,** 17, 1975.

23. Gunnar Myrdal, "Human Dimensions of Development".

24. *Rural Assia: Challenge and Opportunity?,* Asian Development Bank, 1976.

25. *Meeting Basic Needs: Strategies for Eradicating Mass Poverty and Unemployment,* ILO, 1977.

26. T. N. Krishnan, "Demographic Transition in Kerala: Facts and Factors", *Economic and Political Weekly,* **I** (31-35), 1203, 1976. The same point was made a few years ago in a World Bank Report on Colombia: *Economic Growth of Colombia: Problems and Prospects,* World Bank, 1972, p. 308.

27. The Programme of Action adopted by the World Employment Conference, June 1976, also stresses the importance of international economic reforms and co-operation to reinforce action at the national level. Thus it declares that "the satisfaction of basic needs is a national endeavour, but its success depends crucially upon strengthening world peace and disarmament and the establishment of a New International Economic Order". Furthermore, "the Conference recognizes that the basic-needs strategy is only the first phase of the redistributive global growth process". *Meeting Basic Needs; Strategies for Eradicating Mass Poverty and Unemployment,* ILO, 1977.

28. B. T. Feld, Editorial, *Bulletin of Automatic Scientists,* March 1976.

29. "The trading infrastructure (of processing, shipping, credit, distribution, etc.) is largely in the hands of the rich nations at present, so that the poor nations get back only a fraction of the final price that consumers pay for their commodities in the international market. The final consumers pay over $200 billion for the primary exports of the developing countries, these countries receive back only $30 billion." "Proposals for a New International Economic Order," Third World Forum, August 1975. See also A. Salam, *Bulletin of Atomic Scientists,* **32,** (7), 9, 1976.

30. It is significant that a UN Study Group on the Future of the World Economy, under Wassily Leontief, also concludes that, even in order to achieve the relatively low income target of $500 *per capita* by the year 2000, it would be necessary to create especially favourable international conditions relating to trade and payments, technology transfer and technical assistance. *The Future of the World Economy,* Oxford University Press, New York, 1977. The Gabor Report, *Beyond the Age of Waste* (Club of Rome, 1975), also refers to the inadequacy of the present institutional structures and questions the optimistic expectation that problems can be solved within the existing framework of marketing economy.

31. A promising programme of this kind has been going on at the University of Bombay, India. See reference 32.

32. B. M. Udgaonkar, "Science and Technology for Rural Development: Role of Institutions of Higher Learning", COSTED Seminar on Science and Technology for Integrated Rural Development, Madras, March 1977.

33. M. Kaplan, *Pugwash News letter,* **14** (1, 2), 1, 1976.

34. B. M. Udgaonkar, "Development, Resources, Security and New Directions for International Scientific Co-operation: Implications for a Code of Conduct for Scientists", Proceedings of the 26th Pugwash Conference, Muhlhausen, GDR, 26–31 August 1976.

35. China provides an example of a country with rich cultural heritage, trying to depart from the linear model of development and evolve a path of its own. Self-reliance has played a very important role in this development.

36. See, for example, G. S. Aurora and Ward Morehouse, "The Dilemma of Technological Choice in India: the Case of the Small Tractor", *Minerva,* **12,** 433, 1974.

37. For example, Eleanora Masini, reference 4, and references quoted by her. See also B. M. Udgaonkar, references 34 and 48.

38. K. Mushakoji, "An Overview of the Human Social Development Programme", UN University, March 1977.

39. Albert Tévoédjrè, "For a Contract of Solidarity', International Institute of Labour Studies, Geneva 1976; quoted by Elenora Masini, reference 4.

40. The COCOYOC Declaration, adopted by the participants in the UNEP/UNCTAD Symposium on "Patterns of Resource Use, Environment and Development Strategies", COCOYOC, Mexico, 8–12 October 1974.
41. See, for example, B. M. Udgaonkar, "Research and Basic Human Needs—Closing the Widening Gap", UNESCO Seminar on Research and Basic Human Needs, Venice, 1975; *New Quest*, **1**, 31, 1977.
42. D. P. Panchal, "Antenna Development in India: Case History of a Technology Spin-off", *Electronics—Information and Planning*, **3** (11), 873, 1976.
43. In this context one may refer to J. D. Bernal, *Science in History:* "where science has been kept a mystery in the hands of a selected few, it is inevitably linked with the interests of the ruling classes and is cut off from the understanding and inspiration that arise from the needs and capacities of the people."
44. Reference 25, p. 27.
45. The IFIAS programme on a Biological Pathway to Self-reliance and Decentralized Production would be of interest from this point of view.
46. One may also note that there is a certain contradiction between such a model of decentralized development and models based on technology transfer from the ever-growing powerful transnational corporations.
47. Philosophers have been trying this since times immemorial; while their exhortations have influenced individuals, they have had very little effect on the attitude of the society as a whole.
48. A. Forti, Background document for the Venice Seminar on Research and Basic Human Needs, December 1975. See also, B. M. Udgaonkar, Reference 41.

Appendix I

Self-Reliance

The following extract from the Secretary General's Report on the 24th Pugwash Symposium on "The Role of Self-Reliance in Alternative Strategies for Development", held at Dar-es-Salam, Tanzania, 2–6 June 1975, may clarify the concept of self-reliance:

[*Pugwash Newsletter*, **13** (2), 65, 1975]

"SR as a strategy for development involves the following components:

1. SR stresses the achievement of independence, in contrast to dependence, at all levels of the development process. It calls for basic changes in the mode of involvement of developing countries into the world political, economic and cultural system. If the predicament of "helplessness", in which many developing countries find themselves, is to be avoided they must recover the centres of decision-making involving their own destiny.

This process may result, and sometimes has resulted, in a temporary detachment of some countries from the ongoing international economic system, followed by a recoupling into the system on a new basis. The most appropriate strategy for achieving SR must be decided by each country in the context of its particular predicament and needs.

2. SR often takes the form of an exhortation, a political and motivational force, a normative concept which is inevitably ideological in character with a powerful ethical component to it.
3. SR is not necessarily autarchy or self-sufficiency: SR and self-sufficiency do not necessarily lie along the same "line". While SR may often imply or indeed call for self-sufficiency in particular aspects, the reverse is not automatically true; SR is much more than economic nationalism.
4. *What is SR?* Basically, there are two complementary aspects to SR:
   (a) On the decision-making level, SR implies the will to build up and use a capacity for autonomous decision-making and its implementation in all the elements of the development process, including science and technology (S & T):
   (b) With respect to S & T, SR requires the development of an indigenous capacity to generate and put to use those elements of technical knowledge which an autonomous decision-making process has selected for indigenous supply. In particular, it calls for harmonizing technologies of different vintages and capital intensities so as to make the best possible use of the available resources and, at the same time, to engage in a process of technological change, that will result in a continuous increase in productivity. SR development means "walking on several legs".
5. *SR for what?* SR is an essential component of alternative development strategies aimed at guaranteeing the satisfaction of the minimum needs of the entire population and looking toward the setting of ceilings on personal wealth and consumption in both the developing and highly industrialized countries, thus reducing and ultimately removing inequalities. SR is also an essential component in the initiation and execution of an economic growth process of a kind which makes possible the attainment of such minimum needs in a finite time.
6. SR is an essential component, not only of the overall "style" or strategy of "A *New* Development" but also of the actual *processes* by which it is brought about. It takes different forms in different fields—food, energy, finance, technology, etc. Balanced progress towards the desired development objectives may well call for trade-offs between the degree of SR attainable in each of these fields. The central task of executing "The New Development" is to ensure that these trade-offs are made in such a manner that the basic aspects of the self-reliant quality of the development process as a whole (sect. 4) are not compromised.

7. SR is involved in choices made and actions taken at different levels: the individual (self-confidence, fulfilment of personality), village, district, province, country, leading to the collective SR of the developing countries as a whole within the context of an equitable world order.
8. SR calls for participation at all these levels. A close interaction is called for between the various sources of SR; SR can and should apply to the people, at one end of the line, and to the national political leadership at the other, with professionals, such as scientists and technologists, playing potentially important roles in the overall communication and consolidation process involved.
9. *SR and forms of co-operation* Many, if not most of the existing forms of collaboration with "The North" have directly led to the present forms of dependence of "The South". There is great and urgent need to remove the international, political, economic, and cultural obstacles to the attainment of the self-reliant development of the developing countries. These obstacles include dysfunctional "north-copied" values which have been internalized by the leadership groups in many developing countries. The removal of such impediments to the achievement of meaningful SR requires the collaboration among Third World countries for collective (shared) SR. It also calls for the reorientation of the policies, practices, and institutional mechanisms of assistance provided to developing countries by highly industrialized countries, in such a manner as to further the individual and collective SR of the developing world.
10. It was noted in passing that a degree of success in promoting SR in the S & T sphere could be achieved through the selective importation of already-developed technology, provided it is insisted that the national utilization of the imported technology must be accompanied by the provision of indigenous technical personnel and facilities for its independent exploitation—i.e. the example of Brazil. However, most participants felt that this example did not contain enough of the other elements, essential to an SR development programme, to be useful as a model."

Strategies for Change: the Transition

The concepts outlined above will only gain credibility to the extent to which they are translated into concrete strategies for change towards new development styles designed in such a way as to minimize their social cost. The key variables for the transition strategies appear to be:
(a) consumption patterns linked with income distribution and employment patterns; (b) technological styles, translated into guidelines for the evaluation and selection of appropriate technologies; (c) patterns of land and resource use.

Transition strategies may entail partial reconversion of some existing industries towards the production of basic needs-oriented goods; a new pattern of investment priorities—i.e. a different deployment of certain national, regional and world-wide economic activities; an emphasis on employment in socially-oriented services and the promotion, whenever possible, of the exploitation of renewable resources on a sustainable basis in order to harmonize socio-economic and ecological objectives and to promote a greater stake in international economic co-operation on the part of mineral resource-poor developing countries.

Implications for International Action

It was recognized that there are certain actions by the international community which promote the attainment of scientific and technological self-reliance, and other actions which hinder its attainment. It is important to identify these actions and make them explicit.

In the first place there is a set of actions which can be followed by governments of developing countries, either through regional groupings or collectively, which will promote collaboration between developing countries and hence strengthen their collective self-reliance. Several specific suggestions were made about ways in which, this might be done and there was general agreement that, as a minimum, some form of Developing Country Organization should be established. This organization would actively help its member countries in their efforts to attain scientific and technological self-reliance.

Ways in which international organizations and bilateral agencies might promote self reliance were also discussed. One suggestion was that as a policy guideline in considering aid projects, both donor and recipient nations should give highest priority to those projects which contribute most to the attainment of the two components of self-reliance specified in section 4 above. Whereas some international agencies and governments recognize the validity and importance of self-reliance and are already actively helping developing countries to achieve it, many do not, and their "aid" programmes may actually hinder its realization.

SECTION 3

Global, Regional and National Approaches

Research and Human Needs: An Attack on Global Problems

SAM NILSSON*

One characteristic of the modern world is man's increasing ability to anticipate some of the problems inherent in any form of growth and also to change prevailing trends on the basis of such anticipations. Adaptation to future circumstances which can at present only be guessed at may turn out to be the most novel and interesting feature of our time. The quality of human life and of the environment will be determined more and more by deliberate human choices rather than by prevailing conditions and natural forces.

It may be precisely in this circumstance, however, that we will find a dividing line between the industrialized and the less developed countries. The industrialized countries have during the last few decades built up the intellectual, economic, and technological capacity needed to anticipate and to tackle problems by making deliberate choices, while most of the less developed countries still lack these capabilities. For the first time in history, man has the chance of eradicating poverty and human misery for all people on Earth, if his collective intellectual and physical resources can be focused on human needs rather than on destructive capabilities. Science and technology represent a tremendous potential for meeting basic as well as other human needs. In this respect, it is essential that science and technology not be seen as a resource which is utilized only during difficult times; they should be integrated into the socio-cultural development process from the beginning. The 1970s have been a decade of anticipation, alarm and concern, with the United Nations playing a very important role through its conferences on world problems, and many private initiatives helping to increase the awareness of the gravity of the situation facing mankind. The 1980s should become the decade of action and practical achievement.

One of the most important lessons from the big international conferences during the seventies has been that global problems do not necessarily imply global solutions. While the interdependence of all humans and all ecosystems makes it imperative to think globally, in practice each social group must solve its problems locally, even if outside help is needed. We must begin to work for

---

* Dr. Sam Nilsson is Director of the International Federation of Institutes for Advanced Studies, SOLNA, Sweden (IFIAS).

the elimination of concrete evils rather than for the realization of abstract goals. We must not allow our dreams of a beautiful world to lure us away from the claims of those who suffer here and now. The millions of children who are born to this world every week and who soon become subjected to the permanent injustice caused by malnutrition are not much helped by promises of a future "ideal society". The problems and circumstances will change all the time and we must develop institutional mechanisms at the global and local levels which permit quick adaptation to the new challenges.

In this paper an attempt is made to consider research in relation to human needs. It is divided in three main parts: the global dimensions of human needs; localization of human needs; and opportunities for research.

Global Dimensions of Human Needs

A uniform list of human needs cannot be drawn up. Food, housing, clothing, employment, and basic medical facilities are obviously basic needs, but the importance of these needs will, however, change with time in each society and also vary considerably between societies, as S. Radhakrishna of COSTED has pointed out.[1]* Aggregated statistics may tell us little about local human needs, but will indicate the magnitude of the global challenge and the time at our disposal to change the situation. Tables 1 and 2 present figures which show the numbers of people suffering from deficiencies of one sort or another, and the development gap between low-income and high-income countries.

These figures are staggering. But they give us no idea as to how we should go about tackling the problems. Needs are relative and definitions of minimum standards have little practical meaning. To try to determine the level of tolerance of human misery would be futile. After all, people with incomes of the order of $10 a month do keep living, albeit miserably. Moreover, the statistics may be inaccurate and misleading. In a recent research project carried out by IFIAS on Climate Change, Food, and Society,[2] it was found, for example, that the International Classification of Diseases does not reflect the real significance of malnutrition as a cause of death. Often diseases such as measles and dysentery, normally non-fatal, are reported as the causes of death of children.

In the UNESCO document SC–78/WS/54(1978) on Priorities and Trends in Research related to Human Needs, the following categories of needs were discussed: Food/Nutrition, Health, Housing, Clothing, Energy, Education, and Employment. Below is an outline of the global dimensions of some of these needs and the corresponding problems which should be susceptible to research at the international or local level.

---

* Superscript numbers refer to Notes at ends of Chapters.

TABLE 1

In a total world population of 4 billion (1975)		
Undernourished	at least	570 million
Adult illiterate		800 million
Children not enrolled in school		250 million
with no access to effective medical care		1500 million
with less than $90 income per year		1300 million
with a life expectancy of less than 60 years		1700 million
with inadequate housing		1000 million

(*Source:* Center for Integrative Studies, University of Houston, Texas, USA, 1977.)

TABLE 2

Development Gap, by Groups of Countries

	Low-income countries (under $300)	High-income countries (over $2000)
Mid-1978 population (millions)	1331	1102
Average *per capita* GNP	$166	$4976
Average physical quality of life index	40	93
Average birth rate (per 1000)	39	16
Average death rate (per 1000)	16	9
Average life expectancy (years)	48	71
Average infant mortality rate (per 1000 live births)	135	20
Average literacy rate	34%	97%
Average *per capita* education expenditure	$4	$224
Average *per capita* military expenditure	$6	$255

(*Source:* James P. Grant, Overseas Development Council, Washington DC, USA, 1978.)

Food and Nutrition

The provision of sufficient food is the primary basis of individual survival. There is, in fact, enough food in the world today to feed everyone, if the current production were equally distributed and properly stored—which it is not. In the industrialized countries, agricultural productivity is high and can be increased, although at a cost, while population growth (less than 1 per cent) is low. In the Third World countries, on the contrary, agricultural yields and productivity are very low while population increase is more than 2 per cent. It is therefore in the developing countries that a dramatic push is needed to increase agricultural productivity.

The fact that there is starvation and widespread malnutrition today, while food supplies are still ample, demonstrates clearly that the difficulties have been those of distribution economics and the political system. Food is costly to produce; the hungry are poor and unable to buy it. This present situation augurs badly for the future, when the total population will have doubled, and indicates the urgency of preparing now, at the political, economic, organizational and technical levels, for a situation which will inevitably develop.

TABLE 3

World Food Production Summary (million metric tons)

	Total Cereals	Root Crops	Sugar	Total Pulses	Total Nuts	Vegetables & Melons	Fruits	Total Meat	Hen Eggs	Total Milk	Vegetable Oils	Fish[a]
World	1358.8	566.5	81.9	46.8	3.6	294.7	255.6	117.7	23.3	427.2	41.5	50.8
Developed Market Economies	480.2	72.3	24.4	3.4	1.6	94.1	101.9	55.9	11.3	207.5	12.5	20.1
North America	285.2	17.1	6.2	1.1	0.4	27.1	25.5	25.1	4.1	60.3	9.8	2.5
Western Europe	146.9	48.4	12.8	1.8	1.1	49.0	62.7	24.2	5.0	125.4	2.3	7.8
Oceania	18.1	0.9	2.9	0.2	—	1.3	2.2	3.5	0.2	12.8	0.1	0.1
Other	30.0	6.0	2.5	0.3	0.1	16.8	11.5	3.1	2.0	8.9	0.3	9.7
Developing Market Economies	419.6	175.2	40.6	24.2	1.7	101.7	123.7	21.7	3.5	83.1	20.0	14.4
Africa	41.9	74.2	2.8	4.7	0.6	9.9	22.9	3.3	0.5	6.3	3.7	2.0
Latin America	80.2	45.9	24.1	5.0	0.1	13.4	52.7	11.3	1.8	29.0	4.5	0.7
Near East	53.2	5.7	2.5	1.7	0.7	25.6	14.8	2.7	0.4	13.3	1.5	—
Asia & Far East	244.3	47.7	11.0	12.8	0.3	52.4	32.2	4.3	0.9	34.4	10.0	11.7
Other	—	1.6	0.3	—	—	0.3	1.1	0.1	—	0.1	0.3	—
Centrally Planned Economies	459.0	318.9	16.9	19.2	0.3	98.9	30.0	40.1	8.4	136.7	9.0	16.3
Asia	250.6	158.4	4.3	12.3	—	58.6	7.9	16.2	3.8	5.6	4.7	7.6
USSR and Eastern Europe	208.4	160.5	12.6	6.9	0.3	40.3	22.2	23.9	4.6	131.1	4.3	8.7

[a] 1973. From *U.N. Stat. Book*, 1974. estimated proportion of fish catch used for food.
Data: *Monthly Bulletin of Agricultural Economics and Statistics*. April 1976.

The basically political nature of this problem, however, does not excuse us from examining whether limitations of land, water, energy, fertilizers, scientific knowledge and other factors really allow of indefinite expansion without causing major disturbances which would in the end boomerang to prevent the necessary expansion of food supplies. The recent expansion of food production has been impressive; between 1950 and 1970 cereal growth has doubled while population increased by only 50 per cent. But distribution has been very uneven because of the tendency of the rich countries to demand an ever increasing proportion of protein.[3, 4] A recent study[5] carried out in the

TABLE 4

Additional Production required in Calorie-deficient[a] World Regions, 1976
(in thousand metric tons)

Region \ Food Type	Cereals	Starchy Roots	Sugar	Pulses & Nuts	Fruits & Vegetables	Meat & Fish	Eggs & Milk	Fats & Oils
North Africa	826	108	130	68	734	144	287	42
West Africa	780	1373	129	189	961	267	213	42
East Africa	961	570	103	142	536	211	181	31
Mid Africa	478	1077	96	129	739	195	151	32
Mid-South Asia	18,272	7136	1454	2177	7557	2781	2732	495
South-East Asia	6725	2667	553	610	3545	1004	1043	174
Tropical S. America (Less Brazil)	360	286	99	42	430	230	283	35
Total	28,402	13,217	2564	3357	14,502	4832	4890	851

[a] On basis of calorie *per capita* estimates, World Food Conference, *Assessment of World Food Situation—Present and Future,* FAO, 1974.

United States at the initiative of President Carter concludes that the growth in world food production will generally keep pace with the growth in world population, but that in the poorest countries *per capita* food consumption will decline. Moreover, real food prices are expected to increase significantly, whereas they have been declining for decades. The study assumes a 80–90 per cent increase in world food production over the 1970 2000 period, assuming favourable climatic and weather conditions, a continuing successful introduction of high-yielding strains, and major increases in resource use (energy, land, fertilizers) by the agricultural sector. The arable land area is projected by the study to increase by 18 per cent, grain yields per hectare by 70 per cent in reponse to technological improvements and resource use by 180 per cent. This study also indicates that an increasing number of countries will become increasingly dependent on a diminishing number of food exporters. The international grain trade is projected to increase by 180 per cent.

The prospects of developing non-agricultural and non-traditional food sources are not too promising in the short term. It is doubtful if much can be obtained quickly from fisheries whose production has been in decline, although there may be some eventual additions from the catching of species not yet welcomed as food or of krill from the Antarctic, when the whales are no longer there to eat them. Likewise there is no prospect that single-cell proteins, hydroponics or synthetic foods can be produced on a scale to make much difference.

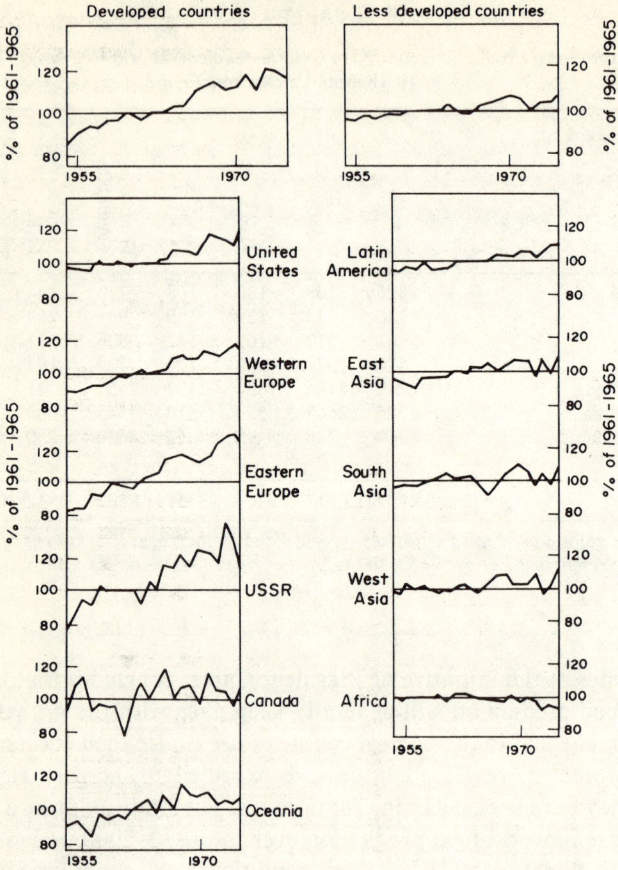

Fig. 1. World food production *per capita,* 1954–1975. The developing countries have gained only 0.4 per cent per year. In none of those regions has the index reached 110, and in Africa it has shown a downward trend since 1951. *Per capita* food production trended upward 1.5 per cent per year in the developed countries until the early 1970s. In each of those regions the index of *per capita* food production has reached or exceeded 110 at least three times in the 22-year period. (Adapted from U.S. Department of Agriculture, "The World Food Situation and Prospects to 1985", Foreign Agricultural Economic Report No. 98, 1974; recent data from WAS–12.)

Very considerable contributions could be made by research into traditional agriculture, with priority to tropical conditions and crops, including some not yet systematically cultivated. But the results of such work, even when successful, can have little influence this century, owing to the time necessary for research and implementation.

Special efforts should be devoted to reducing the energy requirements, including the fertilizer intake, of agriculture. Some promising openings are beginning to appear in the fixation of atmospheric nitrogen.

The growing of legume crops, which have nodules of nitrogen-fixing bacteria in their roots, is the traditional method of introducing biological nitrogen to the soil for use by other crops in a crop rotation system. It is now clear, however, that a much wider range of micro-organisms, both free-living and in various symbiotic relationships with plants, are capable of fixing nitrogen, for example, a recently discovered association between nitrogen-fixing bacteria and tropical grasses and some grain crops such as rice, sorghum, millet and maize, has yet to be sufficiently explored and exploited. Still more distant prospects are held out by the possibility of genetic manipulation of nitrogen fixation elements in various bacteria and plants. Possibilities also exist for producing new nitrogen-fixing plants by tissue-culture techniques, but again I must stress that these are, from a practical point of view, distant prospects.

Considerable advances are possible in the elimination of wastage in storage, particularly in the poorer countries. Overall food losses due to pests such as rats, insects and fungi may amount to 20 per cent or more of the total food supply of the world in any one year, and FAO figures indicate that such pests destroy annually some 33 million tons of food in store, which equals the annual growth in food production. Further invisible losses occur from bad health conditions, common in the poorer countries, where intestinal worms, infantile diarrhoea, and so on, absorb as much as one-third of the food eaten before it can produce nourishment. Many of these losses could be eliminated by better harvesting, improved storage and better hygiene, but this would involve high capital expenditure and better techniques which we cannot expect to be forthcoming very soon.

This suggests that all efforts to meet the food and nutrition needs in LDCs should be directed to production and distribution systems which would increase self-reliance. The concept of self-reliance and its scientific basis has been dealt with in detail in a recent IFIAS book by R. Anderson.[6] Conditions for subsistence farming should be improved by developing agricultural practices which maximize the use of local crops and ecosystems and minimize the dependence on imported resources, such as fertilizers, tractors, and fossil fuels. There is a need to develop agricultural practices which are adapted to tropical conditions and are less vulnerable to drought and other natural disasters. Efforts must also be made to improve sanitary and health standards so that as much as possible of the food produced actually benefits man. Quantitative and qualitative improvements in food production and nutritive standards in LDCs require a great deal of scientific research. This should be a challenge to the entire community and not just to researchers in LDCs.

Land and Vegetation

Is there sufficient land for increasing agricultural production? At first sight prospects seem good, since only 10 per cent of the total land surface of the earth is at present used for food production, and the UN World Plan of Action projection for 1985 suggests a "relatively modest" gain of 600 million hectares or double the area of available land in use in 1962. It is estimated that a quarter of the suitable land in the world, utilized to maximum efficiency, would be able to supply food for several times the present population. But the best land is already in use and as poorer and poorer soils come into use costs increase enormously.

Table 5 shows the usage in absolute and relative terms of arable land for different parts of the world.

TABLE 5

Arable and Cultivated Land and Population, Worldwide

Region	Population in 1975 (millions)	Land area (million km²) Total	Potentially arable	Cultivated	Cultivated area per person (hectares)	Cultivated land as % of potentially arable land
Africa	401	30.2	7.33	1.58	0.39	22
Asia	2255	27.3	6.28	5.18	0.23	82
Australia and New Zealand.	17	8.2	1.54	0.16	0.94	10
Europe	473	4.8	1.74	1.54	0.33	89
North and Central America	316	21.1	4.66	2.39	0.76	51
South America	245	17.5	6.80	0.77	0.31	11
USSR	255	22.3	3.56	2.27	0.89	64
Total	3967	131.5	31.9	13.89	0.35	43

Source: For population: 1975 Population Data Sheet, Population Reference Bureau, Inc., Washington DC. For land: President's Foreign Agricultural Economic Report 298, Government Printing Office, Washington DC, 1974.

The total area of destroyed and degraded soil which has been productive in the past has been estimated at 2 billion hectares, which is 33 per cent higher than the entire area cultivated for agricultural purposes at present, estimated at about 1.5 billion hectares (15 million km²). But the real net loss has not yet been carefully determined and weighed against the potential for increased productivity. This must be an urgent task as a basis for preventing further losses and for creating rational land use practices.

The most obvious option is intensification of agriculture on the existing good lands. According to experts this will necessitate an increase in intensity of land use from 0.4 hectares *per capita,* as at present, to 0.2 hectares *per capita* by the year 2000. Energy analysis suggests that the limit is around 0.1 hectares *per capita,* after which artificial energy substitution becomes prohibitive. There is also a need for reform of the land tenure systems in many countries, consolidation of scattered holdings and the organization of smallholders as a way of achieving greater output. An interesting comparison between high- and low-intensive agriculture for world food production has been made by P. Buringh and H. D. J. van Heemst at the University of Wageningen in the Netherlands.[7] The study suggests that the only way of feeding a world population of 6 billion people without disastrous ecological effects is high-intensive agriculture on the soils which are most productive today. Tables 6 and 7, taken from Buringh and van Heemst, show that modern agriculture on present agricultural land can sustain a world population of 6.673 millions, while labour-intensive agriculture can sustain only 5.356 millions in the best instance.

TABLE 6

Data of Various Systems of Agriculture
(according to P. Buringh and H. D. J. van Heemst)

Agricultural system	Culti-vated land	Harvested cereal crop land	Aver-age yield	Total cons. food	Avail-able food	Popu-lation	Agr. popu-lation
	Mha	Mha	kg-ha^{-1}	Mt	[b]	M	M
Present	1405	928	1,358	1260[a]	300	4200	2000
Modern (on present agr. land)	1406	928	7,287	5338	800	6673	160
Labour-oriented (66)	2462	1625	1,978	1606	300	5353	2000
Labour-oriented (44)	2462	1083	1,978	1071	300	3570	1350
Labour-oriented (33)	2462	812	1,978	803	300	2677	1000

[a] without post harvest losses.
[b] kg *per caput* per year.

TABLE 7

Estimates of Sustainable Population for Different Agricultural Systems
(in millions of persons)
(according to P. Buringh and H. D. J. van Heemst)

	South America	Aus-tralia	Africa	Asia	North America	Europe	World
1. Present population 1977[a]	230	20	410	2400	390	750	4200
2. Modern agr. on present agr. land	474	235	795	3661	890	728	6673
3. Labour-oriented agr. (66) on maximum agr. land	803	190	787	1420	1303	853	5356
4. Labour-oriented agr. (44) on maximum agr. land	535	127	525	947	870	570	3574
5. Labour-oriented agr. (33) on maximum agr. land	401	95	383	710	651	427	2677

[a] Based on World Bank population data (1958) and growth rates.

The "Global 2000" study suggests that growth in arable land area is likely to slow significantly to a level less than half the rate of increase of the past 25 years, despite producer incentives to accelerate the rate of expansion. The slowing down occurs mainly as the result of the shortage of cultivable and reasonably accessible land. The worst shortages will occur in regions with large populations, low incomes, and average caloric consumption levels below the recommended minimum. Because the arable area per person can be expected to decline during the period 1970–2000, major increases in the grain yield per hectare are necessary if individual food consumption is to increase. The improved yields depend on the increasing use of fertilizers, pesticides, high-yielding varieties of grain, and other resource-augmenting factors of production.

Another crucial resource for the preservation of local food producing systems is that of ground vegetation, and particularly forests. The value to human society of healthy forest ecosystems extends far beyond the obvious ones of timber, wood products and recreational resources. A recent World Bank policy paper (*Forestry,* Sector Policy Paper, World Bank, February 1978) states: "The consequences of continued uncontrolled forest exploitation are of critical concern to mankind, for they could lead to serious environmental disruption and increased rural poverty". Over 90 per cent of wood consumed in the developing countries is used as fuel. Over-exploitation of existing fuel wood resources, exacerbated by the recent energy crisis, is leading in many areas to diversion of agricultural residues and dried livestock dung to use for heating and cooking instead of improving soil fertility.

Although the forest area in developing countries exceeds 1000 million hectares, it is being cut down so fast for agricultural purposes that it could disappear within 60 years unless some fundamental changes occur to alter the current trend, or unless extensive reforestation programmes are undertaken to offset the losses. Forests cover one-third of the land area of the world and over half the area of developing countries. Annual world production of forest products is more than $115,000 million; global trade more than $30,000 million. Forests play a significant role in economic development; they provide subsistence, shelter, and employment, as well as resources for the development of other sectors.

But removal of temperate-zone forests also results in heavy losses of nutrients from the soil. Forest ecosystems exert considerable control over patterns of climate, hydrology, circulation of nutrients, erosion, the cleansing functions of air and water, as well as over the status of streams, lakes and underground water supplies. Human activities have already substantially reduced the world's forests and produced a considerable increase in the areas of desert and wasteland. Such destructive activities have accelerated as the human population has increased and, in many cases, complex interactions between human beings and the environment have led to catastrophes. By the year 2000 the balance between forested and cleared land will be greatly altered in Asia, in South America, and in great parts of Africa. Forests will no longer be effective in their stabilizing functions. The consequences of deforestation could eventually become catastrophic. By the year 2000 it may already be too late to restore the environment's former carrying capacity in many of the developing countries.

TABLE 8

Estimates of World Forest Resources, 1978 and 2000
("Global 2000" Study)

	Closed forest (millions of hectares)		Growing stock of commercial sized wood in closed forests and in open woodlands (billions cu m overbark)	
	1978	2000	1978	2000
USSR	785	775	79	77
Europe	140	150	15	13
North America	470	464	58	55
Japan, Australia, New Zealand	69	68	4	4
Subtotal	1464	1457	156	149
Latin America	550	329	94	54
Africa	188	150	39	31
Asian and Pacific LDCs	361	181	38	19
Subtotal (LDCs)	1099	660	171	104
World	2563	2117	327	253
World population (billions)			4.3	6.4
Wood *per capita* (cu m)			76	40

Water

Water is considered by many to be the most critical of all the world's resources, more important even than energy, and may well become the limiting factor for development. Water supply shortages occur throughout the world today, and can be expected to appear with increasing frequency and severity in the future. This is in spite of the fact that the most pessimistic forecast for the year 2000 indicates that less than 50 per cent of total potential water supplies will be withdrawn, and only about 15 per cent actually consumed (that is, unavailable for further use).

The prospect of disabling water shortages in the face of abundant global supplies underlines the intensely local nature of water resources. There exists no world water economy, and it is rarely ever meaningful to speak of a national water economy. The balancing of water supply and demand occurs within smaller hydrologic provinces: single river basins, for example. While water can be moved around and stored as a means of dealing with local supply inadequacies, the necessary infrastructures require relatively large commitments of financial, environmental, and energy resources. They also require long periods of planning and construction, and may interfere with other objectives of the affected societies.

The uses of water cover a broad range of activities. Foremost is the consumption of water for the maintenance of the delicate thermochemical balances within the human body. This in itself amounts to relatively little, for even under extreme conditions of heat, dryness and physical exertion, man seldom needs more than about five litres per day. But as standards of living rise and societies become more complex, the use of water increases. The amount of water used in cities increased by a factor of seven between 1900 and 1960, and by 1980 is expected to reach twelve times the 1960 level (see Table 9). Large amounts of water are diverted for food production—about half the total usage of water today. Water is conveyed not only to intensively cultivated lands under irrigation, but also to extensive areas where partial (supplementary) irrigation may make the difference between good crop yields and practically no yield at all. About 0.5 cubic metres of water is necessary to produce one kilo of dry weight of cereals.

Table 9 shows the water requirements for some products and processes.

TABLE 9

Water Requirements

Use	Amount of water used (m^3)
Drinking water (adult, daily)	0.001
Toilet (1 flush)	0.02
Clothes washer (1 load)	0.17
Refine a ton of petroleum	2–50
Produce a ton of finished steel	6–270
Grow a ton of wheat	300–500
Grow a ton of rice	1500–2000
Produce a ton of milk	10,000
Produce a ton of beef	20,000–50,000

(From P. R. Ehrlich, A. H. Ehrlich, J. P. Holdren, Ecoscience; W. H. Freeman and Comp., 1977)

The common characteristics of water—the large quantities used, and the low user costs—frustrate forecasts of future water use. While it may be tempting to extrapolate from past experience, the possibilities of changes in response to relatively minor adjustments in the way water is managed are so great as to render extrapolation essentially useless. Further problems arise when specific supply forecasts may be compared to specific demand forecasts.

Since water bodies are constantly replenished, the average rate of replenishment is ordinarily of greater interest than the volume of water available at any specific time. Replenishment of surface and ground water resources occurs as a consequence of precipitation. A portion of the total quantity of precipitation is returned to the atmosphere by evaporation and transpiration, a portion becomes inflow to the groundwater reserves, and the remainder runs off as surface drainage. In the absence of human withdrawal, inflows to groundwater resources are usually matched by outflows, through springs or seeps, to surface water bodies. As a first approximation, therefore, the overall rate of replenishment may be obtained by measuring the total surface water discharge from a specific watershed. But unmeasured outflows from groundwater (such as discharges from the ocean floor), net groundwater storage, and/or the existence of groundwater withdrawals may cause this measure to understate the true replenishment rate.

The use of replenishment rate as an estimator of supply carries further liabilities. As the area under study becomes larger (a major river basin or a nation or part thereof), the replenishment rate understates the true supply because it fails to reflect re-use possibilities. As each user withdraws and uses water, wastewater flows are generated and returned to the stream or lake. These return flows are then available to other users, water quality permitting. As a result of these considerations, it is impossible to generalize about the supply of water available to the globe, to any continent, or even to any nation. Meaningful statements describing supply can be made only for relatively small areas, and then after detailed, on-site investigation of the nature and behaviour of the actual water resources available to that area.

Water can, of course, be stored, but storage in reservoirs is becoming more costly. The best available dam sites are already in use, and the remaining locations require heavier investments. But the use of ground water aquifers as storage elements in regional water resources systems may help to solve the problem. The aggregate pore space of aquifers in a region may be several times larger than the available total storage volume on the surface. Aquifers have the additional advantage of not losing water through evaporation, but demand skilled management and input of energy for pumpage.

The connection between water and energy has many aspects. Water is necessary for oil refining and coal mining, and for cooling thermal power plants. Flowing water may have positive energetic aspects (hydro-power generation), or negative (soil erosion and sedimentation). And from here, the link to agriculture and food production is quite clear. We can summarize the problems as follows.

There will apparently be adequate water available on the globe to satisfy aggregate projected water withdrawals in the year 2000; the same finding holds for each of the continents.

Nevertheless, because of the nature of water as a resource, there are likely to be significant water shortages before the year 2000, more frequent and more severe than those experienced today.

Prediction of expected shortage, and the identification of critical water supply issues, requires specific, local, area-by-area analyses of water supply and demand; these analyses are not consistently available throughout the world.

The measures that will be taken to deal with future water withdrawals will have important consequences for the management of the environment and the world's energy resources. Likewise, the development of energy resources, the means employed to manage the environment, and the technologies of future production will have important consequences for the quantity and quality of water required.

Predictions of the nature and the magnitude of these consequences are not available, nor can they be made on the basis of presently available forecasts of water supply and demand.

From this discussion on food and nutrition, land and vegetation, and water resources, a first list of main issues in research can be presented. It is not, of course, exhaustive but may serve as a stimulus for national and international research programmes related to human needs problems.

Issues for Research (Food/Nutrition, Land and Vegetation, Water)

Better understanding of nutrition-performance relations, effects of diet patterns on levels of human functioning.

Mother-child relation regarding nutritional status in different societies.

Nutrition intervention programmes. Better understanding to improve effectiveness in reaching nutritional goals.

Plant breeding and genetic manipulation for high-yield varieties; "classical" genetics; cell biology; genetic pool.

Genetic manipulation for improved adaptability to water salinity increase.

Biological nitrogen fixation specially for grains and other non-legumes.

Increase photosynthetic yields for food plants and trees especially in the tropics.

Improved resistance of major crops to environmental stresses, for example, drought, floods, salinity, acid-deficient soils, and so on.

Better understanding of pest management schemes to reduce pre-harvest losses.

Cultivation of fast-growing nitrogen fixing species of trees for the production of proteins and firewood (e.g. *Leucaena leucocephala;* Casuarina).

Biological control of pests and insect diseases, especially in the tropics.

Effects of large-scale fertilizer and pesticide use on soils and ground water, especially in tropical areas.

Improved usage of tropical soils, better classification schemes of soils, reclamation of degraded soils.

Optimal irrigation and water management schemes adapted to local conditions, especially for small farms.

New fertilizers for tropical conditions.

Better integrated schemes for recycling of agricultural waste, especially for small farms.

Better methods of storage and distribution of crops and of animal food.

Better understanding of breeding, environmental, and management factors affecting aquatic food production.

Better understanding and control of health and production of livestock; of particular importance are foot-and-mouth disease and trypanosomiasis.

Upgrading of traditional methods in agriculture and small-scale production.

Energy and Well-being

Energy is a fundamental input of life, and hence its availability is the key to development. All materials can be regarded as energy in temporarily stable forms, and given sufficient cheap energy the transformation of materials into others should be possible. Therefore (in theory at least) plentiful and cheap energy should obviate material limitations by making substitution possible and should also enable food production to expand almost indefinitely. As long as the sun shines and man is capable of trapping its radiation, energy should be available.

We have become accustomed to consuming the energy of past aeons of solar radiation in the form of fossil fuels, stored in the earth in great but fixed quantities. Until recently these fuels have been available freely and at very low prices. The recent crisis was not really an energy crisis but a petroleum crisis, but it raised the need to develop long-term strategies for world energy use and provision. We have been using cheap and convenient oil and gas for many purposes for which it is by no means essential in the production of low-grade heat for space and water heating, chemical processing and the generation of electricity, for which many other alternatives are more appropriate and less wasteful. As a consequence, much of our apparatus of manufacture is at present geared to the use of petroleum hydrocarbons which are now no longer cheap and, for those industrialized countries which are heavy importers of oil, have considerable uncertainties about continuity of supply.

To gain a perspective of the role energy plays in economic development and the creation of well-being, some of the findings of a recent investigation by Elbek[8] are presented below. In Table 10 are shown primary energy intensity (megajoules per $GNP) values for different sectors of the economy in low-income, middle-income, and high-income countries. The most interesting information from Table 10 is the large difference in total energy intensity between the different country groups, and the distribution of energy intensity over different sectors. Figure 2 shows a diagrammatic presentation of the sector composition of GNP as a function of income *per capita,* based on data for twelve countries from the UN Statistical Year Book.

TABLE 10

Typical Primary Energy Intensity of Sectors (1975)
(B. Elbek, 1979)

Country	*Per capita* 1975 $/yr	Agri-culture MJ/$	Industry MJ/$	Transport MJ/$	Service MJ/$	Households +%	Total MJ/$
Low-income	300	5	100	300	15	9	55
Middle-income	1700	10	150	150	10	10	66
High-income	4800	20	40	100	7	25	33
Asymptotic		16	26	60	6	25	22

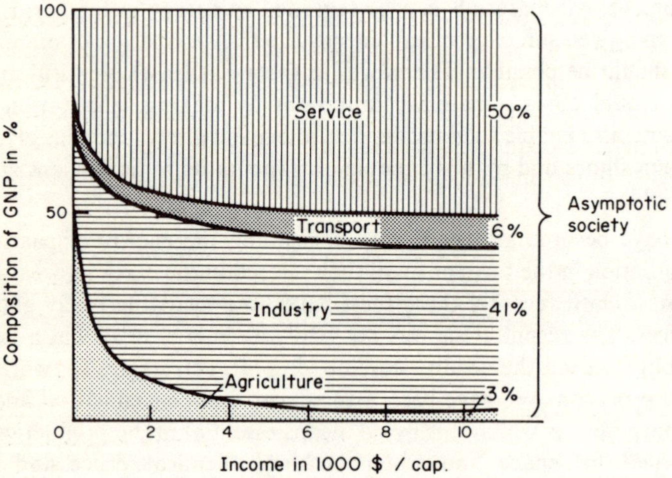

Fig. 2 (B. Elbek, 1979)

From this figure a few interesting conclusions can be drawn. The most significant characteristic is the sharp decline in the contribution made by agriculture, and increases of industry, transport, and services as the GNP *per capita* grows. For the high-income countries, the relative contributions are fairly constant and point to an "asymptotic society" where services account for approximately 50 per cent of GNP. The main reason why high-income countries use so much less energy per unit of GNP is mainly due to modern technology and the large contribution by the less energy intensive service sector.

One of the most important aspects of the energy issue is that of time. In Table 11 the likely energy consumption in less developed countries for 1985 and 2000 is presented. These figures are not linked with development needs, however, Prasad and Reddy[9] have made a calculation of the total energy required to meet economic growth and basic human needs such as food, water, housing, clothing, education, and health care for citizens in India. The figure they suggest is 28,000 kilocalories (kcal) (117 megajoules) per person per day, including both non-commercial and commercial energy. This value is five times the present consumption of the average rural Indian, which may be a very large increase, but it is still very modest compared with current US *per capita* energy use of 202,000 kcal per day. Applying the 28,000 kcal figure to the expected population of LDCs in the year 2000 results in a "need" of 95 MBDOE (209 Exajoules per year, where 1 Exajoule $=10^{18}$ joules). This includes both commercial and non-commercial energy. Today, the LDCs use 18.4 MBDOE of commercial energy alone. There are no reliable estimates of how much non-commercial energy is used in LDCs, but it is probably as high as the commercial energy usage. Assuming that the Indian value given by Prasad and Reddy is a good average for the LDCs one could calculate a current non-commercial energy use around 18 MBDOE. This total is not likely to increase during the rest of this century because population increases are turning forests and pastures into croplands. Hence fire wood and cattle (and cattle dung) may, in fact, decline. If they decline only slightly to 15

TABLE 11

LDC Commercial Energy Consumption
(in MEBOE, million barrels per day oil equivalent)

Country category Source of projection	Base year	1985	2000
Non-OPEC LDCs			
IER[a] (1975)	9.1	—	28.3–41.9
WAES[b] (1972)	9.1	14.9–18.2	26.5–35.6
IEA[c] (1975)	8.9 (1975)	17.5	37.7
OPEC			
WAES	1.7	4.2–4.9	8.4–13.10
IEA	2.1	5.2	14.1
All LDCs			
WAES	9.5	19.10–23.10	34.9–48.7
ODC[d] (1975)	10.5 (1974)	18.5–23.5	41.1–63.6
IEA	11.0	22.7	51.9

[a] Robert Nathans, with Romir Chatterjee, Manuel Taylor and Arandhati Ghash Dastider, *Energy Demand Projections for the Developing Countries* (Stony Brook, New York: The Institute for Energy Research, March 1978).
[b] WAES, *Energy, Global Prospects, 1985–2000.*
[c] International Energy Agency, Standing Group on Long-Term Co-operation, *Technical Report on the Energy Prospects for Developing Countries,* Paris, Feb. 1978.
[d] Energy projections of the Overseas Development Council are included in Hamilton-Rabinovitz, Inc., *Alternative Perspectives on the Economic Evolution of Developing Countries:* Final Report, Prepared for the Agency for International Development, Contract No. AID/Otr C–1570 (Los Angeles; Hamilton-Rabinovitz, Inc., Feb. 1978).

MBDOE, then that figure may be deducted from the 95 MBDOE of total energy to give a commercial energy demand of 80 MBDOE by the year 2000 for the entire Third World. This compares to about 109 MBDOE now being used by the entire world.

Once again it should be pointed out that the 80 MBDOE is not an estimate of what is likely to be used. It is a calculation of what would be required to meet basic human needs if that much could be made available. It is not likely that the target of 28,000 kcal can be achieved by the end of the century. To do so would require that commercial energy use, at present efficiencies, be increased well over four times in a mere 22 years. Such a pace is unlikely to be achieved and therefore there will be a gap between what energy is available and the amount needed to reach the target. A failure to meet basic energy needs for these countries is likely to exacerbate many of the socio-economic problems developing countries now face; these include continuing low levels of agricultural productivity which will reduce the chance of meeting minimal food needs, increased migration to cities (where most commercial energy supplies are available), and a stagnation, at the least, in economic development and programmes for improving the quality of life.

The industrialized countries are forced to make a transition from oil to the sources of energy that will succeed oil during the next 20–30 years. At the same time energy efficiency will have to increase considerably in these countries. The Third World must make the same transition. In so far as their modern sectors are concerned they are at least as oil-dependent as the industrialized countries and hence the transition will be at least as traumatic. But the Third World has another even more difficult transition to make, from traditional energy sources on which most of their people rely to more modern and long-lasting sources. They must make this transition for two reasons: first, because traditional energy is declining—at least on a *per capita,* if not on an absolute basis. Secondly, even if such energy were plentiful it is not of good enough quality to permit these countries to achieve their goals for higher incomes, more food, clean drinking water, housing, and improved education.

In the distance future, the world must approach a state of energy equilibrium. There are several arguments based on resource availability, climate and social needs that makes it plausible that world equilibrium will, or must be reached some time next century. As an ultimate case, let us consider a world of 10 billion people each having an income of $6000 (1975). The resulting energy demand would then be 1300 EJ per year, which is five times that of today, 260 EJ per year. The *per capita* energy use would then be 130 GJ/year, corresponding to 4.1 kW, which is less than the average West European value today. This extrapolation to the equilibrium state is interesting because it illustrates that even a high level of income (West German standard) for a population of 10 billion people does not necessarily lead to excessive energy demands. There are still many uncertainties, however, about the possible impact on the global climate, a problem which is presently drawing the attention of international collaborative research programmes.

It is obvious that the LDCs will have to prepare very careful strategies for the growth of their energy use, which must comprise a mix of several technologies, both commercial and non-commercial. As in the case of agriculture for the satisfaction of future food needs, the LDCs should aim at energy systems that reduce as much as possible the vulnerability of the individual countries to fluctuations in the market place. This suggests optimization and integration of various solar energy systems, both physical and biological.

I list below some of the issues for research in the energy field, which may well be seen as equally pertinent for industrialized countries as for LDCs.

Increase the efficiency (with lower environmental impact) at each stage of the energy system—extraction, transportation, conversion, transmission and end use.

Develop energy systems based on "income energy" that is, solar, wind, geothermal, hydro, ocean thermal gradients, and so on.

Develop energy systems for small-scale, decentralized production of thermal and electrical energy, especially for LDCs (fuel cells, biogas, solar cells).

Determine environmental impacts (local and global) of different energy systems (impact on global climate by CO_2 from fossil fuels, and effects of "energy plantations" on local and global regeneration capacities of biomass and soils, on oxygen and CO_2 cycles, etc.).

Efficient systems for storage of thermal and electrical energy, especially small and medium scale.

Efficient and environmentally sound energy sources for vehicles.

Comparisons of the resilience and vulnerability of different energy systems.

Devise an international framework which will enable all non-nuclear weapons states which want to satisfy some of their energy needs with nuclear power to do so, while minimizing the risks of nuclear weapons proliferation.

Help those developing countries which cannot—or do not want to—embark on a nuclear energy programme to satisfy their energy needs by other alternatives, such as solar, bioconversion, hydropower, geothermal, and coal.

Housing

The possession of a house with adequate light, ventilation and space, affording protection against sun, rain, heat, cold, and such nuisances as insects and other animals, is a basic human need.[10] Shelter, a fundamental human need requiring energy, is often inadequate. Building materials, either recycled or made in small local industries, often deteriorate rapidly. More permanent structures are often costly, and require relatively large amounts of energy to make the materials and operate construction equipment. Hybrid styles, incorporating rural and modern elements are common. Apart from these fixed inputs in construction, there is also an energy requirement for maintenance, and utilities.

The provision of municipal services also requires energy. For example, clean water, essential for health, requires energy to pump it. The World Health Organization estimated that 50 per cent of urban dwellers have no regular water supply and 26 per cent have access only to public standpipes. Meeting water needs will require significant financial and energy investments in pumping, sewage, and treatment facilities.

Fuel for cooking is the major domestic energy need in the slums. With the oil price rise many people were forced to lessen or discontinue the use of bottled gas and kerosene as a cooking fuel and turn to firewood and charcoal. This has increased pressure on forests around metropolitan areas leading to serious deforestation in many places, and subsequently increased wood prices.

Slums are usually the last areas of a city to be electrified. Lighting, radios, TVs and other conveniences, all requiring electricity, are popular demands. Washing and laundry now done without the aid of electrical appliances, usually entail some form of solar energy. Extension of electricity and services to the large numbers of people in slums and squatter areas who lack them would substantially increase energy demands on the cities.

The problem of defining standards for adequate housing is not basically one of research. It is extremely complex and strongly bound to socio-cultural traditions and local life styles which are often extremely difficult to change, even when all material prerequisites are met. One of the principal needs is the development of appropriate technologies which will help create an adequate quantity and quality of housing units that satisfy the human needs of the community. As a result of the UN Habitat Conference in Vancouver in 1976, more comprehensive efforts are being made to define standards and qualifications of adequate housing. The lack of scientific criteria for housing is not surprising, because housing is one of the last areas of human needs to come under scientific investigation. Relevant data are available mainly for the developed countries only. For about 40 per cent of the world's population practically nothing is known and this population is located entirely in the LDCs. From the data available it is obvious that housing conditions, especially in the poorest countries, have worsened rather than improved during the past decade. The average size of a household in Africa, Asia, and Latin America is 6.3 persons. For the developed countries it is roughly the same, 5.7 persons per household. The essential difference lies in the age mixture of the households. The number of younger people per household in the LDCs is about twice the number in the developed countries.

The major need area in the immediate future in Africa and South Asia is for rural housing. In Latin America, where over 60 per cent of the present population is urbanized, the most urgent need is urban housing. Overall, more than 1000 million new dwellings need be built by the year 2000, or 47 million annually.

Issues for Research

Develop cheap methods for the exploitation and use of locally available materials for construction.

Develop concepts and methods of building that are ecologically optimal.

Develop low-cost and non-wasteful techniques for providing water.

Establish minimum standards for shelter, infrastructure and services for dwellings in different climates.

Develop methods and schemes to reduce the migration from rural to urban areas, especially in LDCs.

Develop methods and concepts to conserve energy at the various points of end use in settlements.

Stimulate research on the economic and small-scale energy sources (solar, wind, geothermal, biomass).

Develop low-cost and flexible transportation systems in LDCs.

Health

In recent years many diseases have been eradicated or diminished thanks to improved measures of hygiene, the use of insecticides or the appearance of new drugs. This has, of course, encouraged the population explosion. Nevertheless, poor health continues to be the reality for huge masses of the population. Table 12 shows the magnitude of the major diseases still prevalent in the poor countries.

The state of health of the population of any country is a result of very many factors, not only medical research or conventional health services. The approach to health problems falls into three categories: the curative approach to identifiable disease; attention to poor health that results from such factors as low nutrition, debility, the minor disorders of childhood and chronic conditions of old age; and preventive medicine, including also public health, adequate nutrition and housing, clean water, and a hygienic environment with proper sanitation.

A considerable re-think of health services is in progress in many of the LDCs. More emphasis is put on decentralization of health programmes and the involvement of paramedical personnel. Natural medicines long practised

TABLE 12

Disease	Numbers affected	Disease	Numbers affected
Amboubiasis	Extensive	Plague	Fewer than 1500 cases per year (1960s)
Ascariasis (Lergo Roundworm Infection)	Approx. 25% of world population	Schistosomiasis (Bilharzia)	Approx. 200 million
Bacillary Dysentery	Numbers of cases in the millions		
Cholera	Approx. 250,000 cases (1971–2)	Syphilis	Extensive
Endemic Typhus (lorne)	(1969) 25,000 cases	Trachoma	400 million (1960)
Filariasis	200 million	Trypanosomiasis and Leishmaniasis, and American Trypanosomiasis (Chagas Disease)	Endemic in Central Africa at least 7–10 million in South America
Gonorrhea	65 million new cases a year (1964 est.)	Tuberculosis	15–20 million
Leprosy	11–12 million	Whooping Cough	70 million cases a year
Hookworm Disease	500 million		
Malaria	Approx. 100 million cases each year, 1 million deaths		
Measles	100 million cases each year		
Onchocerciasis	20 million		

Data: G. W. Hunter, ed. *Tropical Medicine,* 5th edition (W. B. Saunders Co., 1976): P. E. Sartwell, ed. *Preventive Medicine and Public Health,* 10th edition, (Appleton-Century Crofts 1973). J. S. McKenzie Pollack, "Planning A Healthier World", Unpublished manuscript. Health Sector Policy Paper, World Bank, March 1975. 1974 Report on the World Social Situation, U.N., Published 1975. Fifth Report on the World Health Situation 1969–1972, W.H.O. World Health, July 1976.

in many LDCs are coming back. Simple ways of protecting and monitoring well-water and other water supplies that provide adequate sanitation and sewage disposal, elementary hygiene in the processing and preparation of food, infant and child care training, and so on, can be more effective than many grandiose and expensive schemes for large medical complexes. A case in point is the tropical disease schistosomiasis, a man-made disease that is probably much more effectively controlled by appropriate water management than by extensive use of molluscocides or vaccines which may eventually be developed.

Preventive medicine and effective therapy of disease depend on advances in medical science. Currently, the world-wide effort in medical research is quite large, being of the order of $5000 million per year. But only a pitifully small proportion of these funds is directed towards the great health scourges of the third world. In particular, tropical communicable diseases continue to take their devastating toll in terms both of mortality and morbidity. A high population of the people of the world suffer from the six most serious of these diseases—malaria, schistosomiasis, typhanosomiasis, filariasis, leprosy and leishmaniasis. The World Health Organization estimates that world research effort on tropical communicable diseases currently totals only 30 million. There is a need for new measures, including vaccines and better drugs, as traditional measures such as vector control have clearly failed. Thus a greatly expanded world-wide effort aimed at control of tropical communicable diseases is urgently required.

Health problems are closely associated with nutrition problems, and again inadequate nutrition, including that at the pre-natal stage, has an important although not fully understood relation to the general quality of development of the individual, for example his capacity to learn. Yet in many places nutritional levels are both quantitatively and qualitatively below those which indigenous foodstuffs, together with plants not yet used for food, could provide. An integrated approach—nutrition, fresh water, housing, medical care and education—must be applied. More emphasis should be put on primary health care methods than on luxury hospitals. Education is an essential element.

Issues for Research

Better understanding of the effects of nutritional standards on fertility and the methods of fertility regulation.

Better understanding of relations between health and nutrition status, specially at foetal and early childhood stages.

Development of methods for self-reliant, decentralized primary health care.

Determine relations between the socio-economic and the health status of people.

Better methods for detecting carcinogenicity and mutagenicity.

Determine the accumulated burden of pollutants on man from air, water and food with special emphasis on populations with low nutritional standards.

Integrated approaches to the control of tropical diseases, especially malaria, schistosomiasis, trypanosomiasis and filariasis.

Establish quantitative relations between health standards and availability of water sanitation.

Education

Education may well be the most important element with regard to facing up to all the difficulties and future needs discussed above. Clearly the right to education is one of the basic human rights, but the question arises as to what kind of education, since it has to serve many purposes, the transfer of the accumulated knowledge of a culture to new generations, awareness of the world as it is, a capacity to learn throughout life, to say nothing of imparting immediately useful vocational skills.

Much effort has gone into the problem of abolishing illiteracy in less developed countries, and this has only partly succeeded, which is verified by the data of Table 13. In many places where a basic educational system teaches the elements of reading and writing, a high proportion of the rural population quickly forgets these arts through lack of opportunity to practice them—absence of books, newspapers and other material. Yet in some of these populations, though technically illiterate, there can be considerable cultural maturity, with people well versed in local political affairs and the events of the world learnt over the radio.

TABLE 13

Adult Illiteracy (15 years and older): 1970

	Number of illiterates (in millions)			Illiterates as % of adult population		
	Male	Female	Total	Male	Female	Total
Developed[a] countries	9	16	27	2.4	4.3	3.5
Developing[b] countries	306	450	756	40.4	60.2	50.2
Total	315	266	783	28	40.3	34.2

[a] All Europe, USSR, USA, Canada, Japan, Israel, Australia, New Zealand and South Africa.
[b] All other countries excluding China, North Korea and North Vietnam.

Data: *1972 UNESCO Yearbook, pp. 47–48.*

Education in the developing countries, while of course it must contain many universal elements, should be shaped as far as possible to fit the local environment and the history, traditions and life styles of local populations. Education is an integral part of the development process; people will not modify existing habits of thought, work methods, or nutritional and hygiene practices unless they understand why, and how they will benefit.

Education and its institutions are generally very traditional, derived from the function of transmitting the accumulated knowledge of the past and the cultural heritage to succeeding generations. Furthermore, teachers usually teach what their teachers taught them. This approach has proved adequate in times of stability and slow change. Today, however, we are witnessing the transformation of the world and of society and it would be a tragedy if the young were to be prepared for life in a world which is disappearing. There is need, therefore, for a new approach to education, an education which accepts and prepares for a changing world and which provides a capacity to adapt and to welcome the challenges of change. The education for the future is likely, therefore, to be different from that of today; less academic perhaps, but aiming, in addition to imparting the essential intellectual skills, at inculcating a capacity to learn and to adapt throughout life. For this we need to understand the learning process much better than we do now. The consequences of educational reform will probably be easily accepted by the flexibility of youth, but they will also have to be assimilated by those already adult, a group which is biologically less flexible. Success in the implementation of development policies, in the assimilation of new technology in hitherto non-industrialized regions, in creating indigenous and creative capacities for research and development, and in many other essentials, will only be achieved if public understanding of the nature of the problems and the goals of the society is profound.

Since education is intimately integrated into all other human needs and closely tied to the local socio-cultural conditions, it is difficult to indicate in a clear-cut way what the main issues for research may be.

Issues for Research

The most pressing need in most of the LDCs is for minimal literacy.

Emphasis has been broadened from reading and writing skills to include "functional" literacy which means giving greater importance to the individual's local experience and needs.

It is important to combine literacy training with other human needs strategies—health care, nutrition, family planning, and housing programmes. The human needs focus might also be generalized for other educational requirements, i.e. at the primary, secondary, and technical levels. This approach meets the more recent criteria for education as a lifelong learning process.

Education and training systems must be developed for promoting scientific literacy among school-going children to enable them to "cope and act" with changes in the world in which they live rather than becoming "master of the facts".

What relations exist between education and employment in different sectors of the economy, i.e. agricultural, medicines, engineering and technology industry, and services?

Problems of meeting needs in secondary, technical and vocational education are different in kind and scope from those of more basic education. Many secondary and upper-level educational services in the LDCs are even more strongly oriented to external models than primary education.

Given the urgent need for skilled manpower and the lack of available resources to meet this need in many LDCs, education is one area in which much increased international co-operation and assistance will be required.

Many LDCs are now endeavouring to expand and improve their primary education as rapidly as possible. Some seek ways by which existing schools can serve by double-shifts, alternating schooldays for different groups and by keeping schools open longer. Others have reorganized the school systems' organizations.

In another mode, satellites are being used more and more for transmitting primary education programmes to remote areas. New types of media should be introduced at a rapid rate, for example, radio, film, TV and cassettes.

Employment

The LDCs have an estimated unemployment total of about 80 million and will have to find about 700 million new jobs between now and the year 2000. According to ILO, jobs will have to be created for almost 1000 million people before the year 2000. Table 14 presents the ILO figures for the total labour force for the period 1960–2000.

In the LDCs, present population structures are expected to result in a relatively greater increase in the work force than in the total of the population. In order to intensify and extend agriculture so as to provide food for the growing population, a considerable proportion of the greater work force will, in any case, have to be absorbed by the agricultural sector. But the main hope of achieving a substantial increase in the living standards of many countries still resides in industrialization. The target of the United Nations—that the industrial production of the less developed countries should reach 20 per cent of the world's total by the year 2000—is probably not realistic, but considerable increases are to be expected, with important repercussions on the markets of the highly industrialized nations. Undoubtedly one of the main objectives of the developing countries will be to produce jobs as well as products, which means that their industrial policies should not automatically emulate the high manpower-productivity approaches of the traditional industrial system.

But the exclusive use of labour-intensive techniques will neither solve the problems of LDCs nor reduce their dependence on industrialized countries. Likewise the exclusive use of capital-intensive techniques will present the LDCs with serious problems: financial difficulties, lack of managerial staff and supervisory personnel and delays in the solution of the employment problems. Thus the LDCs should arrive at a reasonable balance between labour-intensive and more advanced capital-intensive techniques, with the

TABLE 14

Estimates and Projections of Total Labour Force and the Related Growth Rates—World and Regions—1960–2000

Region	Total labour force							Average annual growth rates					
	1960	1965	1970	1975	1980	1990	2000	1960–1965	1965–1970	1970–1975	1975–1980	1980–1990	1990–2000
	millions							per cent					
Developed Market Economies	277.0	295.0	311.9	332.0	351.4	386.1	420.1	1.3	1.1	1.3	1.1	0.9	0.8
North America	79.9	86.8	95.7	104.2	112.6	125.4	139.5	1.7	2.0	1.7	1.6	1.1	1.1
Western Europe	141.2	145.9	147.3	153.1	158.8	170.1	178.1	0.6	0.2	0.8	0.7	0.7	0.5
Oceania	5.0	5.7	6.4	7.1	7.8	9.1	10.7	2.5	2.5	2.0	1.8	1.6	1.6
Other Developed Market Economies	50.9	56.6	62.5	67.6	72.2	81.5	91.9	2.2	2.0	1.6	1.3	1.2	1.2
Eastern Europe and USSR	159.5	166.2	171.8	184.2	194.7	207.7	219.0	0.8	0.7	1.4	1.1	0.6	0.5
All Developed Countries	436.5	461.2	483.7	516.2	546.1	593.8	639.1	1.1	1.0	1.3	1.1	0.8	0.7
Developing Market Economies	519.9	570.2	632.2	705.0	791.9	1010.2	1303.1	1.9	2.1	2.2	2.4	2.5	2.6
Africa	93.7	102.9	113.8	126.1	140.9	179.0	233.8	1.9	2.0	2.1	2.2	2.4	2.7
Latin America	70.8	79.2	89.2	102.0	117.1	155.4	207.3	2.3	2.4	2.7	2.8	2.9	2.9
Near East	45.0	49.5	55.1	61.7	69.7	90.6	119.8	1.9	2.1	2.3	2.4	2.7	2.8
Far East	308.9	336.9	372.4	413.1	462.1	582.3	738.6	1.8	2.0	2.1	2.3	2.3	2.4
Other Developing Market Economies	1.5	1.6	1.8	2.0	2.3	2.9	3.6	1.9	2.1	2.3	2.4	2.3	2.4
Asian Centrally Planned Economies	341.0	363.8	392.7	424.4	456.3	525.9	603.7	1.3	1.5	1.6	1.5	1.4	1.4
All Developing Countries	860.9	934.0	1024.9	1129.4	1248.2	1536.1	1906.8	1.6	1.9	2.0	2.0	2.1	2.2
World	1297.4	1395.2	1508.6	1645.6	1794.3	2129.9	2545.9	1.5	1.6	1.8	1.7	1.7	1.8

Note: The above regional aggregates are derived on the basis of the ILO country by country estimates and projections of the labour force. For more detailed information, please see *Labour Force, 1950–2000,* Vols. I, II, III, IV and V, ILO, Geneva, 1977.

perspective of achieving the fundamental aim of maximizing economic progress and employment and satisfying other human needs. The ultimate purpose should be to increase their self-reliance. Foreign countries and corporations should assist in introducing technologies which are both growth- and employment generating and adapt the technologies to the material and social needs of the host country. Technological and social research should be undertaken within the LDCs with the foremost purpose of meeting human needs. Technicians, professionals and skilled workers should be given assured status and incentives to prevent the brain drain.

For most of the presently industrialized countries, a series of quite different issues is likely to arise. If the present low levels of fertility persist, there will be a considerable reduction in the proportion of the total population in the work force. This may well be welcomed on employment grounds, but will bring its own difficulties and have repercussions on industrial and technological policies, especially for the resource-poor nations. It will mean, for instance, that a much higher proportion of the population will be elderly than at present, which will put great strains on the health and welfare services, resources for which will have to be provided from the work product of the shrinking proportion of active people.

Issues for Research

Development of technologies which are labour-intensive rather than capital-intensive, specially for LDCs.

What socio-economic and political incentives promote full employment (participation) in LDCs, specially in sectors related to the human needs satisfaction?

Optimal utilization of indigenous natural resources to stimulate employment in LDC countries taking external trade into account.

What type of agriculture and industries satisfy best both the employment needs and other human needs such as food, housing and health?

How should women optimally be integrated into the economic and civic life of a country?

Better understanding of the optimal integration of technology from industrialized countries into LDCs to create more jobs.

Interdependence of Problems and Nations

From this survey of human needs the interdependence and interaction of the problems is immediately obvious. Rapid population increase, for example, means a commensurate increase in the need for food, materials, energy and housing. Food production in turn depends on energy to a considerable extent, through the need for fertilizer and farm machinery as well as for the distribution of its products. Energy production, increased use of materials and even agricultural intensification have a marked impact on the environment, which could inhibit agricultural yields. Demographic structures and trends may have an influence on employment out of all proportion to the total population increase in some Third World countries, while in the developed world, they can influence educational planning on the one hand and throw great burdens on the health and welfare services on the other.

Contemporary problems exist, therefore, as an untidy tangle of intertwining difficulties which is sometimes referred to as the "world problematique". Nearly all of them are the direct or indirect consequences of technological development or have important scientific facets. We are facing, not a series of discrete difficulties which can be clearly delineated and tackled successfully one by one but a cluster of interacting problems so tangled the one with the other that it is increasingly difficult to apply solutions without disturbing other areas of the tangle. To attempt to tease out individual strands of the "problematique" seems, therefore, rather like removing the external symptoms of a disease which has not been fully diagnosed. Removal of symptoms may initially appear to be an improvement or even a cure, but it can often change the balance of the system as a whole or cause other symptoms to break out elsewhere in the body.

Recognition of the interdependence of nations, which has been greatly intensified by the advances of technology, necessitates a return to a holistic view of the world, its problems and its needs. In a planet of limited dimensions such as ours, with great unevenness in the distribution of both energy and mineral resources, of climate and of regions suitable for agriculture and the growth of large populations, the interdependence of the individual nations is an inevitable phenomenon. Such interdependence between human societies has, of course, existed for thousands of years, manifested mainly through international trade and political and military alliances. In today's complex world, however, built on technology and demanding large and increasing quantities of materials and energy by no means always available from local sources, all countries in the world (with the possible exception of the largest powers) are increasingly dependent on the others for the maintenance of their well-being.

Economic opportunity ignores national frontiers. The industrial revolution led to the establishment of manufacture close to deposits of coal and iron ore and to the building of great cities on the estuaries of rivers from which industrial products could be distributed throughout the world. This phase is, however, long since passed, and new industries tend to arise, not in proximity to deposits of minerals and fossil fuels, or even on the traditional manufacturing sites, but near to centres of scientific, technological and managerial skills, on the assumption that reasonably cheap raw materials and energy could always be imported. This has become, indeed, one of the facets of interdependence, implicitly accepted as one of the bargaining points of the developing countries in pressing for a New International Economic Order.

What is new in today's world is that interdependence has begun to restrict the power of sovereign governments to guide their own national destiny. The new elements are mainly the vastly increased scale of human activity and of international trade, the rapid rate of change, the growth of expectations on a world scale and the shrinking of the world through technological developments in transport, communications and the effectiveness of the media.

These may not all be sound and healthy trends. There are strong arguments for encouraging and maintaining a diversity of cultures in the world and for considering how development can be achieved in harmony with these. A system of bland uniformity would be a poor substitute for the diversity of cultural systems which has existed in the past, each with its own unique cluster of human values and its own creativity. A world made to a single pattern would be a spiritually poor world and its uniformity could probably be maintained only through controls and dictatorship. From a biological point of view also, diversity would appear to be invaluable in preserving the human capacity of adaptation to changing conditions. But the maintenance of cultural diversity would not mean a retreat into parochialism and mini-nationalism or the preservation or archaic and static social systems. What we envisage is the encouragement or a wide variety of value systems and cultural patterns, interacting and reinforcing one another with a world of mutual interdependence. As far as development is concerned, respect for such a vision would demand a self-reliance approach which blends economic betterment with appreciation of human needs, social and spiritual as well as material.

Localization of Human Needs

From all this it should be clear that the definition of human needs and suggestions for meeting these needs are best made by each individual country. Solutions to the local problems should not be imposed from outside, because both needs and solutions are extremely sensitive to local conditions and cultural values. In the UNESCO document SC–78/WS/54 a conceptual model was suggested that might be used by any local community or nation as a basis for mobilizing research on human needs problems (Fig. 3).

It is important to emphasize that the conceptual model is oriented on "end results" rather than on "inputs". Politically defined **national development goals** in terms of population policy, socio-economic status, and environmental quality should indicate the quantitative and qualitative needs of food, health, housing, energy, education and employment. It is only at this stage that it is possible to estimate the research efforts needed to tackle the problems.

Another concept closely related to self-reliance is that of "carrying capacity". This is a measure of how large a population a country can sustain at a given quality of life within the limits of its human and natural resources, capital and environmental quality. A few countries (Australia, New Zealand and Canada) have begun to apply the "carrying capacity" approach in their long-term development strategies. Few countries—if any—are today completely self-reliant or have a "carrying capacity" that suffices for the demands of their populations. Self-reliance can, of course, be strengthened by exchange and trade with other countries and the notion of collective self-reliance has been introduced recently by some LDCs to indicate that these could reduce their one-sided dependence on industrialized countries by closer mutual co-operation in science, technology, trade and economic development.

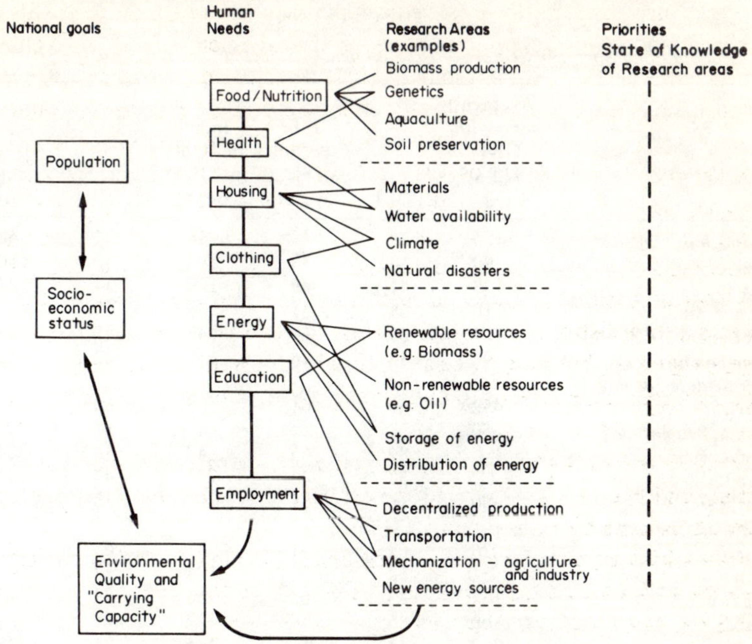

Fig. 3

One of the main objectives of the UNESCO programme "Contribution to the determination of research priorities linked with human needs and societal goals", is to help orientate research and mobilize the scientific community in both rich and poor countries for programmes and problems linked with self-reliant development and human needs satisfaction. A first important step to be taken is a better understanding of the human needs in quantitative terms. Not only the economy based on products and profits has to be expressed statistically, but also the economy based on well-being. We have not yet succeeded in this, but programmes within OECD and UNESCO have been progressing slowly towards the development of social indicators. There is a growing consensus that the traditional measure of a nation's economic progress—the Gross National Product (GNP)—cannot provide any satisfactory estimate of the extent to which human needs are met. Recently, the Overseas Development Council[11] in the USA has suggested a means of calculating a so-called Physical Quality of Life Index (PQLI). This index comprises three indicators—infant mortality, life expectancy at age one, and literacy—into a single composite index having a low of 0 and a high of 100. Table 15 presents the PQLI of different groups of countries together with other socio-economic indicators.

TABLE 15

Economic and Social Indicators of Development in Selected Countries, Early 1970s[a]

	Average *per capita* GNP 1970–75 ($)	PQLI	Life expectancy at birth (years)	Infant mortality per 1000 births	Literacy (%)	Birth rate per 1000
Low-Income countries	156	39	48	136	34	40
Mali	90	14	38	188	5	50
India	133	41	49	129	34	34
Kerala, India	110[b]	68	61	58	60	27
Sri Lanka	179	82	68	47	81	26
Bangladesh	92	32	46	153	22	47
Lower Middle-Income countries	452	60	61	95	60	30
China, People's Rep.	350[b]	68	62	65	60	22
Korea, Rep.	464	82	65	47	88	24
Zambia	415	38	44	159	47	50
Angola	600	16	38	203	13	47
Upper Middle-Income countries	1026	67	61	87	65	36
Iran	1260	52	57	104	37	45
Mexico	996	75	65	66	74	42
Cuba	800[b]	85	70	27	78	21
Algeria	780	41	53	145	26	48
Taiwan	847	87	70	25	85	26
High-Income countries	5272	93	71	17	96	17
Kuwait	13,787	75	69	44	55	43
United States	7024	95	73	15	99	15
Netherlands	5558	96	74	11	98	13

[a] Income categories include only the 127 countries listed in the World Bank's "Derived Atlas Series".
[b] Single-year estimates for 1975.

Sources: GNP data are in constant 1974–76 dollars and are from World Bank, "Derived Atlas Series", November 1977, except for China, Cuba and Kerala. The GNP data for China and Cuba are from Population Reference Bureau, "1977 World Population Data Sheet". The data for Kerala are from John W. Sewell and the Staff of the Overseas Development Council, *The United States and World Development: Agenda 1977* (Washington, D.C.: Overseas Development Council, 1977). The social indicators are based on data from Population Reference Bureau, "1978 World Population Data Sheet". Literacy data are from *UNESCO Yearbook*, 1973 and 1975.

A major advantage of consolidating the three indicators mentioned in the PQLI is that such a composite index summarizes a great deal of social performance. Use of life expectancy or infant mortality alone, for example, often leads to the mistaken conclusion that these problems can be solved by the medical practitioners alone. Together, however, the three indicators provide important information about how the benefits of development are distributed. They also capture the effects of social, economic and cultural policy far more accurately than is possible through analysis of income distribution alone. The results of any research effort—whether national or international—linked with human needs and societal goals, should be measured in these broad terms.

A second operational issue raised by the notion of meeting human needs is that of the goals to be achieved. It is in the setting of targets that national and global development efforts may be brought to converge. Professor Jan Tinbergen[12] has called for the attainment of the following national objectives for all countries by the end of this century:

A life expectancy of 65 years or more (compared with low-income countries' present average of 48).

A literacy rate of at least 75 per cent (compared with 34 per cent in low-income countries today).

An infant mortality rate of 50 or less per 1000 births (compared with the present average of 136 per 1000 in the low-income countries).

A birth rate of 25 or less per 1000 population (compared with the present average of 40 per 1000 for low-income countries).

The setting of goals and the development of indicators which measure the quality of life, may become the most constructive policy tool for guiding national and international research efforts as suggested by the conceptual model of Fig. 3. The most important characteristic of the model is that it is multi-disciplinary and interactive. The opportunities for research in sharpening the policy tools are many—research that must be carried out both at the national and international levels.

The Scientific and Technological Potential of LDCs

A basic assumption in the presentation above has been that science and technology can contribute to the solution of human needs problems. To be really effective, science and technology must become integrated elements in national development plans. This leads us to examine the present scientific and technological potential in LDCs. It has been estimated that only around 5 per cent of the world's research and development is carried out in the LDCs, and in the majority of the LDCs only an insignificant fraction of the national resources is devoted to science and technology. Although admirable efforts are being made there is a risk that the intense day-to-day pressures will close future options or delay indefinitely the creation of an indigenous and sustainable basis for the systematic use of research in meeting human needs.

The international scientific community has a great responsibility to help the LDCs out of this dilemma by reorientating some of their own research in the direction of human needs problems, and by making available to the LDCs the scientific and technological information relevant to human needs problems. In addition, private and intergovernmental organizations should assist in building up local capabilities of science and technology. There is no easy path to the creation of a vigorous scientific and technological capacity. It must be approached simultaneously from many angles—in the universities and technical colleges, in agricultural institutes and extension services, in industry and in the public services. This building-up process necessarily takes a long time and cannot succeed unless there is general understanding among politicians, public servants, the business community and the academics of the essential need to create a national infrastructure, coupled with the other policies and programmes of the nation. The creation of such a capacity for science and technology, together with an understanding of the nature of the technological innovation process, must be the initial and major elements of a science policy of those countries in early stages of development. The ability of the LDCs—both individually and jointly—to create, choose, adapt, manage and utilize science and technology, is fundamental to their prospects for sustained and self-reliant economic and social progress.

References

1. *Research and Human Needs—Problems of Developing Countries,* S. Radhakrishna, COSTED, March 1979.
2. *Drought and Man: The 1972 Case History,* an IFIAS project. To be published by Pergamon Press.
3. *The Two Faces of Malnutrition,* Erik Eckholm and Frank Record, World Watch Institute Report No. 9.
4. *Picture of Health, Environmental Sources of Disease,* Erik Eckholm, Norton Inc., New York, 1977.
5. *Global 2000,* ed. Gerald O. Barney. To be published by Pergamon Press.
6. *Biological Paths to Self-Reliance,* Russell E. Anderson, Van Nostrand Reinhold, New York, 1979.
7. *An Estimation of World Food Production Based on Labour-Oriented Agriculture,* P. Buringh and H. D. J. van Heemst, Centre for World Food Market Research, Wageningen, The Netherlands, August 1977.
8. *World Energy Perspectives.* Bengt Elbek, Niels Bohr Institute, University of Copenhagen, Denmark, 1979.
9. K. K. Prasad and A. K. N. Reddy, "Technological Alternatives and the Indian Energy Crisis", *Economic and Political Weekly,* August 1977.
10. *Ten Basic Human Needs and their Satisfaction,* Sarvodaya Development Education Institute, Moraturva, Sri Lanka, 1978.
11. *Disparity Reduction Rates in Social Indicators,* James P. Grant, Overseas Development Council. Monograph No. 11, September 1978.
12. *Reshaping the International Order,* Elsevier, Amsterdam, 1976.

The Research and Human Needs Programme in Latin America

OSCAR NUDLER

The Research and Human Needs programme was launched in Latin America in 1977 with a meeting organized by the Fundación Bariloche of Argentina. The meeting was held in Bariloche between 30 May and 1 June 1977, and was attended by twenty-eight researchers from various areas connected with the human needs field. It was decided at this meeting that the Fundación Bariloche would coordinate the programme and a series of recommendations for future action was adopted, one of which was "to form a coordinated network of persons and institutions of the region who would work on the subject of human needs".

As a first step it was decided to hold another meeting the following year. It was agreed that it would be a seminar to discuss papers specially prepared for the occasion, within the following agenda:

(a) Theoretical and empirical studies on human needs in general.
(b) Human needs in the context of Latin America.
(c) Prospective studies on quality of life.

The proposed meeting took place between 25 and 28 June 1978, at the headquarters of the Centro Latinoamericano de Economía Humana, Montevideo, Uruguay. Immediately afterwards and with the same participants, a meeting sponsored by the Society for International Development was held on the Subject "Human Needs and the Development of Latin America". In view of the similarity of the topics, this second meeting was practically a continuation of the first, so that a total of 5 days of valuable discussion on human needs and development was achieved. Forty-three participants attended, most of them from Latin America. Numerous papers which had previously been distributed to the participants were discussed. Some of those papers have recently been published by the host institution in a book entitled *Research and Human Needs in a Development Strategy for Latin America*. This publication received the support of the Division of Scientific Research and Higher Education and of the UNESCO Regional Office of Science and Technology for Latin America and the Caribbean.

The third Latin American meeting of the Research and Human Needs programme took place in Tiradentes, Brazil, from 25 to 27 October 1979. The agenda of this meeting, devoted to "Human Needs and Childhood" included subjects such as society and the needs of childhood, the role of the child in different development styles, childhood and poverty in Latin America, childhood and health, and so on. Since this Brazil meeting, special attention has been paid in the Latin American programme to the needs of underprivileged, the relatively sectors of the population such as children, women and ethnic and other minorities. Following this line, the fourth meeting—held from 9 to 12 September 1980—was dedicated to the problems of women in so far as they relate to the basic concerns of the Research and Human Needs programme. This meeting was held in Mexico City and the host institution was the Centro de Estudios del Tercer Mundo. Papers were presented and discussed on women and family in poor and marginal sectors, women and urban and rural work, women, marriage, children, feminist movements in Latin America, women's vision of the future, and research priorities and the problems of women. The Proceedings of the third and fourth meetings are to be published both in Spanish and English, this latter as part of a joint collection of books by Fundación Bariloche and Pergamon Press.

As a result of the Brazil meeting, a permanent consulting board composed of people from different Latin American countries was appointed. In the Mexico meeting this consulting board was converted into a steering committee and an executive secretariat was created. The ultimate objective of this structure is to accelerate the building of a permanent research, communication and action network throughout Latin America. After 4 years of fruitful work and growing influence, the time is ripe for reaching this stage. It is hoped that on the primary basis of the effort of the Latin American institutions and persons involved and with the co-operation of UNESCO and other international agencies, that goal will be fully attained in the very near future.

Future Developments

A first step that has already been taken towards shaping the new phase of permanent, collaborative undertakings of the Research and Human Needs network in Latin America was to devise some research projects. Three of these projects are considered as having first priority: underprivileged children and women in Latin America; human needs research as a key for building up a human development theory applicable to the Latin American context; building up a Research and Human Needs reference centre.

To illustrate the conceptual basis of these projects I shall quote from the introduction to Project 1, Underprivileged Children and Women in Latin America:

> "This research project focuses on the situation of underprivileged children and women in Latin America, starting from a human needs

point of view. The term 'human needs' is interpreted here in a very comprehensive sense. Even if survival or 'basic' needs have no doubt first priority, all human needs comprise an interconnected system so that the permanent or prolonged dissatisfaction of one part of the system is bound to affect the whole. This systemic approach to human needs leads to conceiving the scope of the project as broader than that of the more usual sort of studies dealing with social indicators of poverty, such as low levels of nutrition, health, housing and so forth. Without denying the very first priority of analysing and confronting the grave problems included under those headings, the approach followed here embodies the assumption that human beings, whatever the conditions in which they live, cannot and should not be split or dichotomized between material and non-material sides, or between lower and higher needs. The socio-economic conditions are not seen here as an isolated object of research standing by itself but as a part—a very fundamental part—of a major whole including them together with the political, religious, psychosocial and other conditions which surround people's lives. This also holds in the case of the large numbers of the Latin American population living at a critical subsistence level. It is therefore necessary, when dealing with the physical aspects of poverty, to deal simultaneously with other kinds of poverties, such as the poverties in participation, stimulation, affection and dignity. To put it very shortly, the central idea of this project is to reach a deep understanding of the whole world of the underprivileged and, on this basis, of their needs, satisfactions and frustrations.

"To achieve that end, a far-reaching change in research priorities and, indeed, in the entire current research philosophy is required. We are thinking not only in terms of interdisciplinary co-operation but also in terms of transdisciplinarity. This involves a shift from the discipline-centred outlook into a problem-centred one, as well as an unprejudiced view of the paths leading to knowledge. Our purpose is to use critically as many sources of information and insight as possible, ranging from economic and social statistics up to anthropological and even literary descriptions of everyday life. This does not amount, of course, to a meaningless piling up of information but is intended to imply a carefully designed research strategy for capturing the multi-dimensionality of reality."

The objectives of the project are summarized as follows:

1. The world and life experience of underprivileged women and children in Latin America: and integrated understanding and assessment from a human needs point of view.

2. The role of underprivileged children and women in alternative development styles. Their vision of the future.
3. Scientific research methodology in connection with the study of the problems of the poor and in the light of a comprehensive human needs approach.
4. Study, recommendations and, whenever possible, implementation of concrete actions aimed at radically changing the situation of the underprivileged strata of the Latin American population.

In addition to the joint research projects just referred to, other lines of future action are envisaged. One of these lines is obviously the continuation of the programme of yearly seminars mentioned in the first part of this note, a programme which has proved to be so highly successful in intellectual and human terms. The 1981 seminar will be devoted to the problems of cultural identity, the clash between cultures in Latin America being the focal point. The need for cultural identity is so pressing in the Latin American context, especially in the case of the socially deprived and culturally denied Indian population, that the subject fits quite naturally into our programme.

A third line of future action aims to strengthen the impact of the programme not only on the scientific community, which is already significant, but also on institutions presently engaged in human needs satisfaction, on policy-planners and makers and on the general public. A first concrete step in this direction will be the publication of a book which will explain the scope and objectives of the Research and Human Needs programme in Latin America. This book will, it is hoped, be a useful tool for increasing general knowledge about the programme and, in this way, the influence of the human needs approach.

As well as the book, other forms of presentation of the programme are planned, such as films, tapes and so on. In fact, some first attempts with this purpose have already been made. In the last meeting of the programme in Mexico, in September 1980, a short film on the problems of peasants who migrate to Mexico City was presented by staff from CEESTEM, the host institution. Shortly afterwards, a TV tape was recorded in Argentina on the basis of a play by an Argentine writer and a debate on the characters of the play, seen from the perspective of human needs and human development. This work was jointly supported by the UN University and Fundación Bariloche.

All the future lines of action briefly described here are not mere ideas but real possibilities which, provided the necessary funding is given, will turn into concrete realities. This promising future is mainly grounded on the appeal of these simple and far-reaching concepts of human needs and human development. They endow the programme with an internal expanding force. We have found that they have a remarkable power of unifying around them people coming from different disciplines, holding opposite points of view in other domains, belonging to diverse parts of the social structure and living in distant regions of this large sub-continent called Latin America. We hope the human needs approach to science and to the other aspects of social life and development in Latin America will continue to grow in the years to come. The Research and Human Needs programme of UNESCO is certainly trying to contribute in that direction.

Science, Technology and a New Model of Human Needs

ANDRÉ DANZIN

To link research with human needs implies two premises. First, that needs can be defined and put in order of importance; and second, that it is possible to devise a research policy capable of leading by rational steps to solutions to these needs. This book demonstrates both the difficulties of this procedure, and the desperate need to succeed.

There is no question that the results of scientific research can be used to solve some critical problems and that the diffusion of scientific and technological information will contribute to a better understanding of the problems the developing countries face, and make explicit the conditions under which progress can be made. Producing a catalogue of human needs, however, amounts to the same thing as building a model of a society and then trying to make it real. And who today would dare to propose a universal model suitable for tomorrow's world?

Within another 50 years, 8000 million people will be living in an environment one cannot predict, created by the pressures of technologies yet unknown. That is what we have to prepare for.

Since a universal model is impossible, one possible approach is to look for regional models, different both in their objectives and in their methods, as a way of taking account of the different cultural forces and climatic, geographical and geological conditions peculiar to each major region. There is some wisdom in this approach, but it can lead to confusion. How can Asia find from within her own culture—from Confucianism, from Buddhism or from Islam—practical solutions to the problems of the population explosion or to the impact of new technologies? How can Africa find in the concept of "Negritude" an answer to the spread of the deserts, or the unbalanced terms of trade with the rest of the world? Can Marxism, which arrived on the scene when industrial evolution was already well advanced, take up the problems of complexity, interdependence and diversity which the modern world presents? Certainly philosophy and tradition can help in building mental models of society and in defining human needs, but these forces, representing as they do a mixture of the rational and the irrational, are unlikely in themselves to be able to handle the complex problems man has to solve, more especially as some of those problems are so urgent.

These difficulties explain why the apparent success of the West, and above all of the United States, exerts such a hold on the developing countries. To match or exceed the Gross National Product of the United States: has that not been the objective, explicit or implied, of many governments? But today one thing is quite certain: the American model cannot be extended to the rest of the world. It is quite impossible to provide for every person alive today the amounts of energy consumed by every American, equivalent to 11 tons of crude oil per person per year. And even the European level of consumption, though less than half that of the United States, is equally impracticable. It is not that one can show that fossil and nuclear fuels, solar energy and possibly nuclear fusion would never be able to provide these levels of energy production one day. But it is impossible to image an energy industry capable of providing 80,000 million tons of oil equivalent—or even 40,000 million tons—each day by the year 2030, when the world population will reach 8000 million. There just isn't time, because it takes too long to bring energy resources into production, and the capital investments necessary are not available.

But over and above this purely physical constraint, there are other reasons for considering industrial society as no more than a transition solution, unacceptable in the long term. Industrial society has, in effect, limited the human being to just two roles, producer and consumer. And while both these roles may be essential ones, they are far from satisfying all man's most basic needs. They ignore the fact that man also needs joy, knowledge, affection, an intellectual life, friends and an easy conscience. One index sums up this society: the measurement of GNP. Contemplation of the GNP has become an obsession, and its growth is included as a priority objective in all society's activities. But in fact the GNP is really no more than its originators intended it to be, a remarkable indicator of economic performance, and no proper measure of human contentment and well-being. The GNP ignores, because it has no power to appreciate, most of the pleasanter aspects of human life: anything that is home-made, in the workshop of the garden; the task of looking after the family; cultural activities; voluntary work—without even touching on the "black economy" which has sprung up in certain countries as if in response to the general will.

The English economist Colin Clark calculated in 1956 that measuring an economy by its GNP ignores about 30 per cent of the activity going on in a country like the United Kingdom, and probably double that in the Third World. The part that is neglected, wealth produced outside organized economic structures, is particularly precious because it is the product of individual free will. But industrial civilization, both socialist and capitalist, has valued it so little that it has become one of the aims of the women's movement to gain an equal share of paid employment with men.

Signs that industrial society has not adjusted to the present world situation are multiplying. The system has not reflected the fact that primary resources are limited, as is land for agriculture and housing; nor has it been able to tackle the inequalities of society, or problems arising from the combination of political and economic factors. Unemployment, inflation, monetary crises, violence and the unpleasantness of the huge urban sprawls are all examples. At the same time, it is impossible to take seriously the philosophies and ideologies which have been around for some time.

In this unpleasant situation, the most useful attitude may be to look around for the hidden forces which are already at work bringing about change. What are the factors determining the future? What are the characteristics of the technological society which has been created by the unprecedented success of science? Are we in a period of stability, or only a short-lived state of transition? These are important questions for the three-quarters of the world which is in the process of development, as well as those which have already secured it, because a new model of society implies changes in human needs as well as in the methods for satisfying them. In particular, one can ask the fundamental question of whether poverty can be eliminated by taking a short-cut and avoiding the route followed by today's wasteful and unequal societies.

In what follows we are going to try to assemble some information on the factors of change which one can identify in the most advanced societies. In this analysis we will try to be as cool and objective as entomologists studying a colony of insects. Without pretending to completeness, one can take the list that follows as very characteristic. There will, of course, be many references to energy, since it was the energy crisis which first revealed the instabilities threatening our society.

The Evolution of "Job Analysis"

An American research worker, Marc Porat, working at Massachusetts Institute of Technology at the beginning of the 1970s, had the idea of looking anew at the American employment statistics. He was not interested in the traditional division of manpower between agriculture, industry and the services; instead, he looked for those engaged in the collection, transmission and processing of data.

In a rather arbitrary way, but one which does not diminish the force of his conclusions, Porat divided the working population of America into two groups—those who are in direct contact with products or services in the course of their work, and those who arrange, classify and spread information in the form of figures and symbols. In the first group went all those who work in agriculture in the fields, in hot-houses or in the meat and dairy industry, but not their book-keepers and advisers. These, together with teachers, research workers and those in insurance, marketing, engineering, holy orders or the law, as well as many other categories, were listed and classified as "information workers". These men and women operated without ever touching the things they were dealing with. A bus driver, in direct contact with his bus and with his passengers, would come into the first category, but an employee of a travel agency which arranged hotel reservations and plane tickets clearly would fall into the second. After preparing the lists of these two types of occupation, Porat studied how the relative sizes of the two groups had changed over the past 100 years in the United States. The result he ended up with is rich in implications. In less than a single generation—our generation—those engaged in information have become the larger of the two groups. Today they amount to half the working population, ten times more than agricultural workers, more than twice as many as industrial workers.

Nothing could give a clearer illustration of the evolution of post-industrial society; no index is more sensitive than this one is to the stage of development of a society. North America can show 50 per cent, Western Europe around 40 per cent, Eastern Europe including the USSR certainly less than 25 per cent, while the Third World is lower now than the United States was in 1900, at just over 5 per cent.

Further studies have been carried out, particularly in France, with the aim of separating out from manufacturing proper all work connected with design, job-description, writing safety instructions, installation and maintenance. When these are added to the specialists listed by Porat one can show that in 1979 54 per cent of the French population was involved in the processing or diffusion of information.

The Development of Software

The new word *telematique* first coined in the Nora-Minc report[1]* is intended to include all those activities previously described by the words electronics, telecommunications and *informatique*. Apart from a few details, concerned with printing and photographic techniques, *telematique* coincides more or less with what is called information technology in the United States.

The transmission of messages, words, pictures and diagrams by telegraph, telephone, radio or TV, and the processing of information by computers would all have been understood 20 years ago in terms of the machines needed to do them. One would have understood that it would be necessary to construct transmitters and receivers, lay telephone wires or install aerials and coaxial cables, and build memories and logical circuits for processing data; but one would not have realized that the essence of these technologies was not material at all, but abstract. This was not clearly demonstrated until the beginning of the 1960s when the terms hardware and software were first coined—later translated into French as *materiel* and *logiciel*. Today software

---

* Superscript numbers refer to Notes at end of chapters.

tends to be the more expensive of the two in both the creation and use of information technology. Figure 2 shows how the proportions have changed in the past and extrapolates the curve into the future.

And what is true for data processing tends also to become true for electronics and for telecommunications—that is, for the whole field of information technology. The design of devices and the writing of programs become a major part of making the technology work. In this sense, manipulating the materials has been eclipsed by the labour of the intellect.

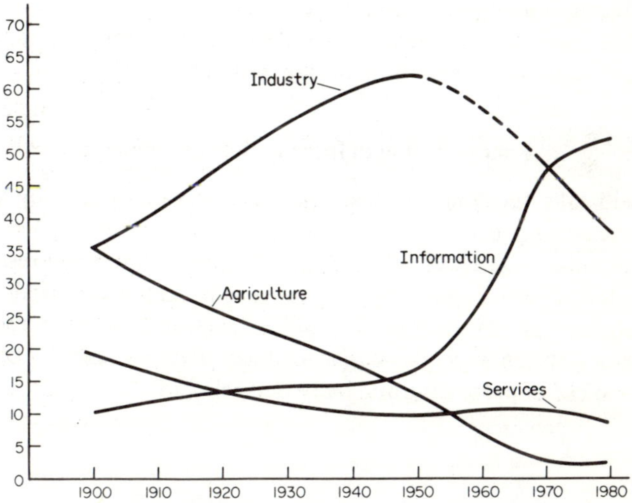

Fig. 1
The distribution of the American labour force, showing the growth of the "information professions"

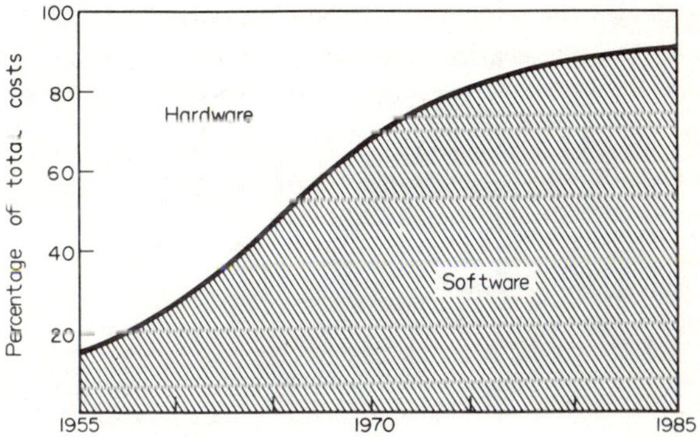

Fig. 2
The split between hardware and software in total costs

These industries are becoming more and more important. If one adds up all the activity involved in manufacturing and servicing the technology of information, one finds that it forms a growing fraction of the GNP. From nothing in the middle of this century, it now represents more than 5 per cent in countries like the United States and Japan. Reasonable predictions suggest that it will amount to 7–10 per cent of GNP by the end of the century in the most highly developed economies. So we are witnessing the birth of a new industry whose principal resource is intellect. Look again at Figs. 1 and 2. Is what is happening in information technology not a foretaste of what will happen to all human work? The shift of the bulk of the work force towards the processing of information (Fig. 1) leads us to think that the concept of software is about to express itself throughout human life (Fig. 2).

Energy Saving in Information Technologies

It would take too long to recall all the twists of fortune by which data transmission and processing has grown in the past 30 years, while reducing the amount of energy it consumes by a factor of 10,000. I will restrict myself to two graphs (Figs. 3 and 4) taken from a remarkable unpublished study by G. Broussaud.[2] It is important to note in passing that these reductions in energy consumption were not sought for themselves but came about for other reasons—speed of processing and increased reliability.

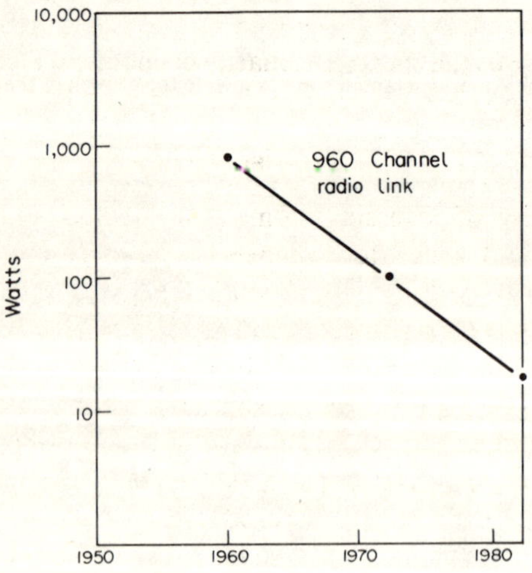

Fig. 3

The reduction of energy consumed by data transmission over 30 years

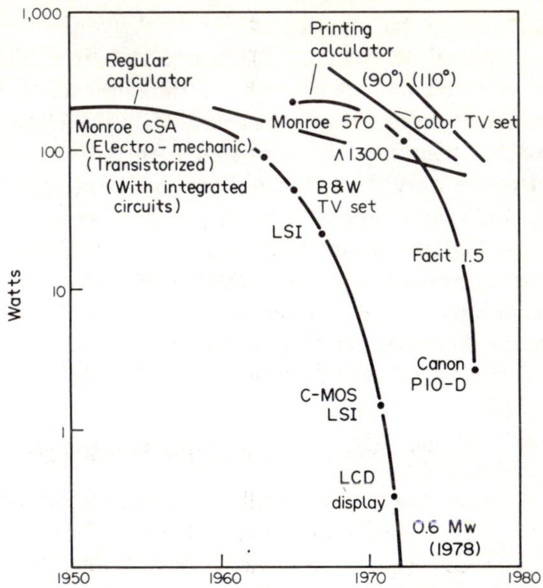

Fig. 4
Reduction in energy consumption of various electronic devices

There is another example of extraordinary energy economy, in biology. In processing genetic, metabolic and growth control information, biological systems are 10,000 to 10 million times more economical than the best electronic circuits in the laboratories. Perhaps scientific research can help us one day make use of these extraordinarily economical techniques.

The Use of Time

The usual way of assessing human progress is to add up the wealth produced—or, more accurately, the wealth which can be counted in the form of the GNP, disregarding unpaid work carried out for the benefit of the individual or of society as a whole. But as far as measuring the quality of life is concerned, this is a poor indicator by comparison with the analysis of how time is spent. What kind of balance-sheet can we prepare of the time allotted to us? How can we set down the amounts of time we spend on our education, on our self-improvement, or on our thought? What proportion of our time do we give to rest, to salaried and productive employment, to transport, to hobbies, to sport, to pastimes, television, do-it-yourself, to voluntary work? If we analyse our time balance-sheet, we will know better how we use both our material and spiritual resources, and so may be able to come closer to understanding human needs.

Among the changes one can identify since the beginning of the century, three are particularly important. They are the time taken for average journeys within the big cities, free time, and the time spent being educated.

Average journey times for salary earners living in large industrial cities have reached 2½ hours a day, compared with only 1 hour a day at the beginning of the century. In an entire life the average citizen, assuming he lives for a normal span of about 73 years, will have spent the equivalent of 6 years queuing, in public transport, or in traffic jams. This is a huge waste of the most precious resource we have, free time, as well as an enormous waste of energy. Such are the facts of life today. New forms of transport which could handle the increased demand are not here yet, so that in these affluent days the average speed of cars in cities is no greater than that of horse-drawn vehicles a century ago. In addition, cities have spread and become divided into residential and industrial zones, generally a long way from one another, just as if the idea was to increase travel times.

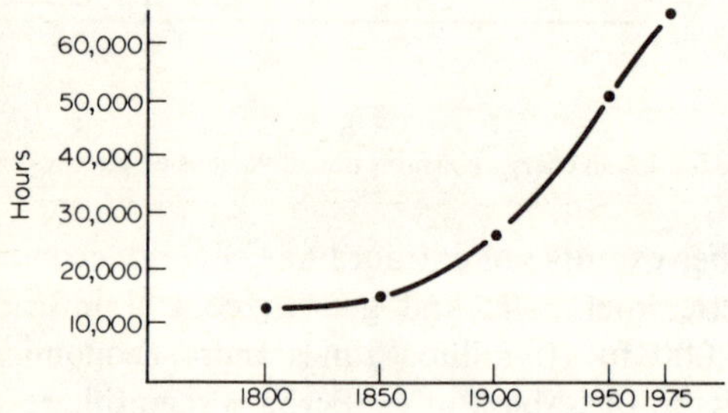

Fig. 5
Total time devoted to travel by people living in cities

The Growth of Free Time

Free time is what is left over after one has subtracted all the other activities: the period of childhood and at school, the time spent sleeping, resting and eating and the time spent working and travelling. What is left, men and women can use as they please; a period for enjoyment, for individual expression and for generosity in taking part in voluntary work. Human nature being what it is, society will probably change in such a way as to increase free time, to the point where it may pose major problems.

Figure 6 shows, for France, how free time has changed over the past 150 years. The curve looks rather like that for a "change of phase" to use an image familiar to physical chemists. This growth of free time has been caused principally by the great growth of productivity, first of agriculture, then of industry, and, in the near future, of the "tertiary" professions by the improvements brought about by information technology.

The problem of free time is redoubled today by the growth of unemployment in the developed countries and by population growth in the developing countries. In the OECD countries, 6–7 per cent of the population is out of work. Either the problems of unemployment can be eased by work-sharing (in which case everybody's free time will increase) or one cannot avoid having large numbers wholly unemployed, for whom the whole time is free. As for the poor countries, how can they hope to find the capital necessary to create jobs? From now on China will be facing an annual demand for 10 million new jobs; Algeria, following the trend of births, is getting ready to face a proportionately even greater problem, with 500,000 new jobs to find every year in the relatively near future. And what can one say of countries like Egypt, India or Mexico?

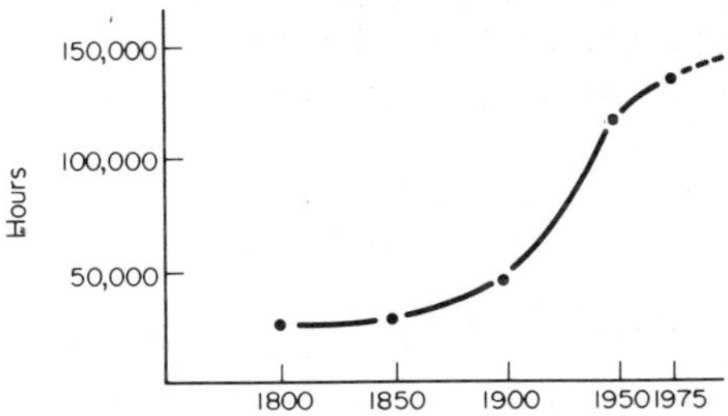

Fig. 6
Total "free time" (in hours) available to a person born in France over his entire life

It is becoming clear, as part of the analysis which the energy crisis has obliged us to undertake, that free time cannot be spent simply doing nothing. That would make the economic problems worse, as well as threatening political stability; one cannot imagine a society which would willingly submerge itself in idleness.

There is, thus, a double interaction between technology and the use of time. The way in which time is divided up establishes the pattern of consumption and, hence, the needs that are felt. And, looking at it the other way, the amount of energy and money available for job-creation establishes also the free time available, as well as its social and geographical distribution. One thing can be taken as very probable: in the most advanced countries, there is unlikely to be a halt or even a slowing down in the adoption of automation, that is to say in the productivity of labour.[3] The strength of the international competition for jobs and the promise of the new technologies are too precious to be abandoned. Besides, if this kind of progress is really possible, why should man want to escape from the prospect of a golden age which would consist of watching machines doing his work for him?

TABLE 1

Products of the Telecommunication, Electronic and Information Industries

Telegraph, telephone, telex
Conference call, facsimile transmission, electronic mail, videophone
Data transmission equipment
Radio and television transmitters and receivers, car radios, radio-telephones
Hi-fi equipment
Photographic and sound studio equipment
Tape-recorders
Discs, audio-cassettes, video-cassettes, magnetic tapes
Radar and radio-detection apparatus
Military electronics
Naval and aerial navigation instruments
Radio-control of missiles and satellites
Scientific instruments (electron microscopes . . .)
Automation and control equipment
X-ray, gamma-ray, infra-red and ultra-violet generators
Robots
Ultrasound generators and detectors
Particle accelerators and detectors
Radio-telescopes
Hand-held calculators
Computers
Computer networks
Computer terminals: screens, printers, graph-plotters
Remote access terminals
Punched cards, magnetic cards, memory cards
Electronic components
Microprocessors and memories
Recorders of temperature, pressure, movement . . .
Lasers
Wires and cables, optical fibres
Telecommunication satellites, weather satellites, spies-in-the-sky
Information systems for documentation, administration, control, economic forecasting, computer-aided design
Electronic clocks

The Growth of the Time taken up by Education

Since the 19th century, attendance at school has been compulsory for all in the developed countries. In practice, the time spent in study has steadily got longer, and the burden of the curriculum steadily increased. Higher education has steadily attracted more and more students, to the point where there is no longer any apparent link between the number of degrees awarded and the availability of high-level jobs.

In spite of these efforts, one can see problems on at last three fronts. It is becoming impossible to pass on the sheer quantity of knowledge science has accumulated; and the content of courses is as often as not out of date, because the professors teach what they themselves learned 10 or 20 years before, to men and women who will face responsibilities in 10–20 years time in a very different technological and political setting. The child and the adolescent have two sources of information, the school and the media, particularly television. These two sources of information often tell a different story.

New technology explodes on to the scene as yet another source of change. To resolve real problems, it is desirable that science and technology be supported; but then new applications of science and technology emerge, tending to disrupt society even further. So long as the concept persists of a foundation of knowledge which must at all costs be passed on to children at the start of their lives, then the problems of the out-of-date curriculum will become more and more insoluble. One should accept, instead, that learning should become something that is done throughout life, and that what one acquires as a child is useful principally as a means of "learning to learn". The use of knowledge and the attempts to improve it and bring it up to date will last throughout life; school, higher education and information will provide a kind of continuum. In addition, preparation for a sensible use of free time is becoming as important as preparation for professional life; the child is to be not only a worker but also a citizen, and must learn how to take control of his own life.

Nothing I have said about the use of time, about the wastefulness of transport, or about the economic, human and educational value of free time is new. Since the 1960s numerous sociologists and philosophers have tried to draw attention to these problems,[4] but it was only when the energy crisis intervened and opened people's eyes that the media began to take an interest in these subjects and political thought began to absorb their message.

The new technologies of information can probably help in solving some of these problems. Through telecommunications linked to data-banks, it is becoming possible to provide those who live in remote country districts with new intellectual possibilities. Bureaucrats will be able to carry out secretarial, administrative or accounting functions from home, just as some so-called "liberal" professions have been able to do for a long time. And if men and women still prefer to work away from home, either because they enjoy the social contact of an office or because they want to make use of the very best equipment available, then the same data processing and transmission technology can be used to locate the offices a long way from the factory and close to where the office workers live. Changes of this sort are already happening in many industrialized countries; an example can be found at Marnes-la-Vallée in France. The whole question of the location of housing and jobs will have to be reconsidered.

Equally, it is to be hoped that we can little by little learn to use for educational purposes the powerful information systems and audio-visual techniques that are now available—despite all the disappointments experienced so far.

The Birth of a Wired Society

The linking of economic and cultural systems across national frontiers is nothing new. History is full of examples: the "tin route" fed the Bronze Age, the "silk route" provided the West with both the riches and the culture of the East. But never, until the past few years, has the traffic across the frontiers reached such an intensity as to change its whole nature.

Air travel and, more important, new information technology have profoundly altered the process of commercial, financial, scientific, technological and cultural exchange. The volume of scientific information published is growing by about 14 per cent a year and is now so large that any single year between 1985 and 1990 will produce a greater number—if not quality—of published works than were produced by all the scholars from the origins of science until the middle of the 20th century.

This huge volume of information is produced in order to be spread throughout the world for the benefit of the scientific community. Thanks to communications satellites, distance makes no difference to the cost, quality, or speed of transmission. Thanks to data banks and computer networks, the information can be made easily and rapidly accessible. The same astonishing developments in information technology can also be found in financial and commercial centres, travel agencies, and in police headquarters for use against fraud and terrorism. Table 2 lists some of these international information links which have become a sort of nervous system spreading through the world, more or less independent of political control.

TABLE 2

International Data Flows

International commercial transactions
Management of multinational enterprises
Remote data processing using specialized systems
Access to large data banks for scientific information, patents, and economic and social statistics
Access to mathematical models for decision-making (economic forecasting, studies of the energy market, projections of population)
Ticket and travel reservations
Transport of freight
Financial transactions (money markets, etc.)
Use of observation satellites, particularly in surveillance and weather forecasting
International surveillance of criminals (drugs, terrorism)
Management of troops by the large military alliances

This list is not, of course, exhaustive. Each of the activities indicated has shown a growth in the volume of information of from 10 to 25 per cent a year. The US Federal Communications Commission estimates the annual growth of international data flows from the United States on leased lines at 21 per cent. For France, the Ministry of Industry predicts growth at the rate of 25–30 per cent a year until 1985.

The effects on international relations may be considerable. Information flows are often the forerunners of more tangible forms of trade, or of the establishment of cultural links. Europe and Japan can only survive through such links, on which neither side can exert real political control. The economies of the Western world and of the Third World are developed through these linkages, within the context of a fully developed free market; each nation has thus become dependent on the world which surrounds it. Only the superpowers, the United States and the Soviet Union, could survive in isolation without suffering great damage.

Countries as different as Argentina and Poland have been able to develop their economies only through the involvement of foreign capital. In West Germany or France every other worker is making products for export. Whole regions of Africa and Asia remain fed only by the import of agricultural commodities. Even China has been obliged to bring her policy of isolation to an end.

This conclusion has very important implications for research into the global needs of mankind. It might be disastrous, for example, if decisions were to be taken for reasons of narrow nationalism. One might imagine a country seeking energy self-sufficiency would think it sensible to replace fields planted with cereals with solar collectors, or with energy "crops". But if the same thing were to be done on a very wide scale, the effects on the world's food supply could be tragic. In the same way, any restrictions on free trade could cause considerable damage, particularly in Western Europe, Japan and South-East Asia.

But the development of the world's "nervous system" by the growing flows of information is far from having only negative effects. Thanks to them, international trade can continue to grow even when people cannot be moved quickly from place to place. They make possible energy savings by allowing certain regions to concentrate on products for which they are better equipped. And in addition, the growing wealth which is one result of the flows leads to the establishment of peaceful relations. As Samuel Pisar has observed, a war between France and Germany would today be unthinkable, not only for pragmatic or humanitarian reasons, but also because the German and French economies have become complementary; one does not wish ill of a neighbour who is one's most active supplier as well as one's most reliable and wealthy customer. The political structure based on this network of exchanges is a lofty conception; the stakes are no longer an area of disputed ground, but technological superiority and the ability to do business. The struggle goes on, but now it has moved from the military plane to the economic and, by cause and effect, to the scientific and technological. Without being a perfect answer, this is progress when compared to the traditional recourse of armed domination for settling disputes.

Towards a New Society

From all this we may conclude that in all probability the advanced countries are in the midst of a new type of development, driven by science and technology. If this is true, it is likely to lead to great changes in the pattern of consumption, and it will require us to think again about time spent in activities other than paid productive employment. As far as simple measures of productivity are concerned, this new society will be less productive. (In this respect it is possible to view the recent decline in productivity witnessed in the United States as less a failure than, perhaps, a sign of progress.) In such a society, one may hope, power will be less tightly organized, more decentralized. The growth of the cities will diminish.

Of course, it would be nice if we could guess what this "knowledge society" or perhaps "communication society" is going to be like. In particular, what would be the world demand for energy, given societies of this sort? It would be equally useful to know the likely outcome of the inevitable struggle for power. Far more than by the ownership of property or the possession of political legitimacy, power in this society will be determined by the ownership of information and the ability to broadcast it. As we have already hinted, struggles may not take place at the military level, but we have no reason to suppose that the normal mixture of human aggression and competition will stop leading to conflicts. New types of conflict will emerge, but in what form? Will there be a kind of diffused violence added to the normal competition for information and know-how produced by the growth of research and development? One area of competition will be in the collection, processing and diffusion of information; to some degree, communication satellites, computer networks and databanks may come to replace coal and steel as the sinews of power.

Can one plan this change? Can one keep it as a general objective and then try to direct it in ways particularly adapted to human needs? Can one hope to propose this "low-energy society" to people in the course of development? This fundamental question is likely to remain open for a number of years yet. Some people, basing their arguments on some observations in biology, hold that human evolution rests on luck; man can only adapt himself as best he can to the determinism of science and technology, much as he would like to determine his own fate. Others have a much more voluntarist outlook and think it possible, and thus necessary, to shape events. Without choosing between these two schools of thought, I would like to try to explain how it is that science has brought us to the point where we are asking ourselves these questions.

These observations illustrate the theoretical ideas discussed in the first part of this book. Industrial society is behind us. It was followed by a technological society which cannot be considered to be a model valid for all time, but only a very interesting period of transition. We seem now to be moving towards a "communication society". Will it be really a new civilization, where knowledge holds the key and science and society are reconciled?

The two soft technologies, of information and biology, can play their part as "triggers" able to turn society towards a new formula in which information, communication and knowledge will be the most important products. In looking back to work out how we got here, one can see a lot of unexpected detours and realize that the unexpected and the imponderables played just as important a part in technological progress as did deliberate effort. And at the end of the day it has all worked out as was hoped by the humanists of the past, who wished for a solid material comfort for man, certainly, but also the advent of the spirit.

From this perspective, one can identify new ways of tackling the two principal obstacles against which technological society has foundered: the threat of scarcity and the smothering effect of growing complexity. There are no scarcities in a civilization whose needs are spiritual, because there are no limits to intellectual, aesthetic, cultural and social activity. As for the damaging effects of complexity, one can legitimately hope that they may yield to the power of data processing and distribution.

A society of this sort will have to respond to the needs of 8,000 to 10,000 million people. To bring it about deserves the efforts of all, but it is clear that the principal responsibility will belong to the rich who alone are in a position to make the transition quickly. It will then be their task to show to the less fortunate a route which alone offers us the hope of breaking the endless cycle of inequality.

References

1. *Information Science and Society*, a report to the President of the French Republic, by Nora and Minc (La Documentation Française, Paris, 1977).
2. *On the Evolution of Telecommunications up to 1990*, by G. Broussaud.
3. This claim is open to the objection that productivity has in fact declined in the United States in the past few years. In fact gains have continued to be made in agriculture, industry, and in the services, but the growing complexity of society which has produced a growth in activities one might call "social software" which has hidden the other gains.
4. *Arcadia, an Essay on the Good Life*, by Bertrand de Jouvenal (Futuribles, SEDEIS, Paris, 1969).

Research and Human Needs: the African Perspective

HENRI HOGBE-NLEND

The purpose of this article is to present the African view of UNESCO's programme on Research and Human Needs. It divides into two sections: first, determining what the most important needs of African societies are, and secondly, choosing priority areas of research with a view to satisfying those needs. My conclusions are based on the results of the inaugural meeting of the African group on Research and Human Needs held at Yaounde, in Cameroon, between the 5th and the 7th of December 1977 under UNESCO's auspices, and on the preparatory work and the results of the conference of Heads of Government of the Organization for African Unity (OAU) on the 28th and 29th of April 1980. This conference adopted a Plan of Action for the Economic, Social and Cultural Development of Africa, known as the Lagos Plan of Action.

The essential characteristics of research as it has developed in Africa in the two decades of independence is that it is a colonial, outward-looking kind of research, whose aims and preoccupations have been directed at the needs of the former colonial powers; both the research institutions and the scientific community itself have been dominated by non-Africans. This is evident not only in university work, but more significantly also in applied research, particularly in agriculture and medicine. The result is that Africa faces, first and foremost, the problem of redirecting research to make it correspond to the real needs of the African people.

The first and absolutely the most important of these needs is the need for development, both at the individual and collective levels. This cannot any longer be thought of simply in terms of economic growth; it should actually be a gradual process of transformation of African society with the ultimate aim of satisfying the needs of the great majority of the population, now and in the future, and of improving their standard of living, both material and spiritual. Taking one's cue from the Plan of Lagos, such a process of development would be collective, rapid, self-reliant, endogenous, and based on the co-operation between African countries and their gradual economic, social and cultural integration. The aim should be to create, between now and the year 2000, an African Economic Community.

At the present stage of development, the most urgent needs of the great majority of the African population are *primary* needs, both material and cultural. By primary needs I mean those elementary human entitlements without which man cannot rise above a primitive level, indeed "cannot live" in the full sense that term signifies or ought to signify in the last quarter of the 20th century; instead he vegetates, he simply exists. These absolutely basic needs include the following:

A balanced and sensible diet.

Basic public health standards.

Decent clothing and housing.

Universal elementary education.

While recognizing that these elementary needs take priority, we should not underestimate the importance of the many cultural and spiritual needs of the African people and African society. Among these, one might raise the following:

African cultural identity.

Inter-African co-operation.

Basic legal rights.

The need for an African cultural identity is one of the most important. Becoming aware of and acknowledging this identity would liberate the African from all sorts of constraints which enslave him. They slow down his development and misdirect it; cramp his style, erode his dignity, discourage him from becoming master of his fate. In one word, they prevent him from becoming a "creator". It is because of this that one must reject so-called "development models" imported from abroad, and the delusive concept of "technology transfer" as a route for the scientific and technological development of our continent.

The experience of African countries in the past two decades shows that the two fundamental ways of controlling science and technology and putting them to the service of development are, first, to create, reassess, develop and use indigenous scientific and technical knowledge, and secondly, to master non-African science and technology so as to assimilate them without being assimilated *by* them.

Africa must follow both these routes simultaneously. The first, on its own, would lead Africa backwards, a senseless course; the second would lead to servile mimicry and the loss of African identity: we must "march on both feet", but not on all fours!

As far as importing technology from abroad is concerned, we should remember that the evolution of science and its applications depends on economic, social, cultural and historial conditions. Its development is neither neutral nor it is complete. African conditions call for new solutions, in the search for which Africans can and must play a leading part. That is why I believe that the idea of "technology transfer", at least as it is generally understood today, is not an acceptable strategy for the development of science and technology in Africa. That can only be based on the development of indigenous creativity. In the final analysis, to be developed is to be a searcher, a creator.

Inter-African co-operation is a pressing need. The balkanization of Africa by the colonial powers means that no African country in isolation (not even Nigeria) can build up the "critical mass" of research workers needed for creating the science and technology needed for national development. The only way to succeed is to combine a continental strategy with regional and sub-regional approaches, establishing how to use scientific teams. It was this concept of a collective and integrated development of Africa which was decided upon by the Lagos Summit. At that meeting, an appeal was made to African countries to abandon the colonial situation in which discussions between North and South are carried on between the former colonial powers and individual African countries, to the detriment of inter-African co-operation.

Africa also needs basic legal rights. That means the establishment of the rule of law, and the preparation and application of legislation adapted to African conditions together with legal rules, either national or covering the whole continent, which favour the growth of individual liberties. This is linked to the decision of the OAU to draw up an "African Charter of Human Rights".

Research in Africa, whether basic or applied, must be directed towards the material, cultural and spiritual needs of African societies. From this point of view, the most important research areas are as follows:

Food and tropical agriculture: including malnutrition (particularly among children and mothers), plants for arid and semi-arid regions, concentrating on food crops rather than cash crops for export, the water resources of tropical soils, techniques for producing and storing food, intensive animal production, the management of teams of oxen, animal health with a view to more intensive production methods, fish farming, forestry research (surveys, management, use of secondary wood, and timber industries) social and economic studies of the production, marketing and consumption of products, and research into the preservation of the ecosystem.

Basic medical needs: including research into medicinal plants, traditional medicine, childbirth, ways of obtaining medicines more cheaply, simple public health techniques for rural areas, simplifying methods of working in country health centres, health education, the promotion of preventive medicine, and of ante and post-natal and pediatric medicine (including the problems of malnutrition).

Natural resources and the environment: including research on the subsoil, mining, new and existing energy sources, ground water and the problems of drought.

African civilizations: research which will restore self-confidence and pride to African man; African history, language, writing, law, arts, literature and technology, plus an effort to make more generally available what is already known.

Science teaching in Africa: with a view to developing a scientific and enterprising spirit among young Africans, to adapting science teaching to the African setting; research into the use of African languages in science teaching.

Industrial research: various industries, including food, forestry, chemical, textile, industrialized building, transport, communications and housing suited to the African environment; research into the adaptation of materials to the tropics, corrosion, industrial pollution and industrial management.

Demography: including research into population growth, settlements, the exodus from the countryside, emigration.

Trade: particularly the deterioration in the terms of trade brought about by low prices for primary materials and high prices for industrial goods and oil.

War and peace: including particularly research into problems capable of starting wars, for example the frontiers established by the colonial powers.

Social research: into the relationships and communication between the different social classes in Africa, both within countries and between them.

UNESCO's programme Research and Human Needs has as one of its aims the idea of being both multi-disciplinary and trans-disciplinary. As such it offers a new and untried framework within which these problems can be studied; for they are, essentially, problems which transcend any single discipline. Our continent is woefully short of such a framework for thought and action.

Problems of Developing Countries: Some Case Studies

S. RADHAKRISHNA*

The problem of providing basic human needs to all is most acute in the developing countries. Ironically it is in those countries that little effort is made to relate scientific research with human needs. Undoubtedly science and technology have helped many developing countries to leapfrog into the modern era in certain areas. Countries like India and Indonesia use satellites for communication, and many people in the urban areas watch spectacular scientific achievements like men walking on the moon on their drawing-room TV brought to them via satellite communication. Countries like Pakistan have advanced colour TV and some African universities have sophisticated science equipment. Yet it is in these very countries that nearly 50 per cent of the people live below the poverty line, suffer from slow and painful diseases, lead a life which is often characterized as living death. They keep crying for food, and begging has almost become a common profession. It is again ironical that in these countries, where scientists, doctors, and technical men probably have the most vital role to play, many such qualified men are underemployed or unemployed and are frustrated.

The picture appears gloomy, but there is hope. The hope lies in the proper application of science and technology. If only well-trained manpower can be diverted to tackle the problem of providing basic human needs to the poorer sections of society, the developing countries have a bright future. Many national organizations, with international support, have been attempting to adopt crash programmes and are spending enormous amounts of money. In one Indian state alone (with a population of 5 million people) nearly 500 million rupees have been allotted to a nutritional programme. If indeed all this effort is being made it is hard to explain why we have not been able to touch even the fringe of the problem. An analysis of the programmes undertaken so far will be extremely useful in drawing lessons for the future. It is also important to interest scientists in developed countries in the problems of developing countries so that they too can use their resources and talents for tackling the problems. It is most encouraging to see that many organizations

*Dr. S. Radhakrishna is Scientific Secretary of the Committee on Science and Technology in Developing Countries (COSTED) of the International Council of Scientific Unions (ICSU).

in developed countries are making special attempts to educate their people about the problems of developing countries. In fact it is strange to note that developing countries are not putting in as much effort in this area. Even while attempts are being made to relate scientific research to human needs several international agencies including UNESCO have been analysing the wide range of programmes that are being undertaken. This report is prepared with the objective of sharing the experiences of COSTED in rural development programmes.

What are Basic Human Needs?

It is often argued that if we attempt to provide food, shelter, clothing and basic medical facilities all the problems of poor people can be solved. While these do constitute basic human needs they do not exhaust the list. Several aid programmes within a nation and between nations have not yielded significant results because they failed to recognize some missing aspects of basic human needs. In the long run it does great damage if food, clothing, shelter and medicines are doled out as charity to poor people. In many development programmes in India, housing settlements were constructed and provided for poor communities. Very quickly it was found that the people sell these houses or rent them out, and return to their own surroundings. Obviously minimum housing standards as laid down according to international standards are not necessarily valid in the developing countries. Obviously there are other more urgent basic needs—but what are these and how can we identify them? In India and other even poorer countries, clothes are often donated to poor people. Very often these are sold in return for money, or more often food.

All this made us believe that priorities for human needs have to be re-investigated and in fact special attempts must be made to identify the "missing aspect". With this objective in mind we participated in some rural development programmes. Very quickly it was realized that no development process will be lasting if the people—the local people—are not involved in every step. They must plan their own priorities and their own schedules based on local needs and aspirations. Our experience has shown that this will lead to a more satisfying development. In certain areas priority was given to health facilities, in other areas the priority was to provide an occupation for rural women during their free time, in other areas it was adult education, in yet other places the emphasis was on improving public sanitation. What was treated as top priority in one place gets the lowest priority in another. No single priority list can be drawn up. The most important conclusion that we would like to draw is that basic human needs are *felt* needs and at least till we gain the confidence of the poorer sections, we must learn to satisfy their felt needs, even if we feel that this is not the best approach. Once their felt needs are satisfied, and confidence gained one can attempt to bring about an improvement in other aspects. The satisfaction of needs should essentially be guided by the following steps.

Bridging the Confidence Gap

Many small farmers lack knowledge of innovative agricultural programmes, methods of controlling pests, or post-harvest technology. They suffer considerable loss and cannot really make agriculture provide them with a living. This makes their life one of abject poverty leading to poor health. The first task should therefore be to make the individual a better farmer. He should be led to gain confidence in his profession and his traditional practices strengthened when possible, or changed by demonstration where necessary. Any attempt to change his basic life-style will only lead to more frustration, particularly because results will not come quickly. Similarly many people in small rural communities are weavers who produce cloth. Their technology is poor, their marketing strategies unproductive and their financial resources meagre. Any attempt to help them meet these challenges will build-up confidence and will bring about new hope. Several such strategies have been tried out in COSTED rural development programmes where the emphasis has been to make a farmer a better farmer, a weaver a better weaver, a tailor a better tailor, and so on. After this initial bridging of the confidence gap the people in a rural area will be open to suggestions about better sanitation and improving the general environment, will be receptive to adult education programmes and will play a positive role in building up their own society.

Basic Medical Amenities

There is another category of human needs—medical facilities—which cannot wait for such patient handling. It is well known and widely recognized that many diseases which afflict rural people are preventable. Our field surveys and case studies have revealed several examples, like hook worm, scabies, child blindness, protein—calorie deficiency leading to Kwashiorkor and/or Marasmus.[1*] Several of these appear because of poverty, insanitary living conditions and ignorance. Many of them can be tackled by concerted efforts which do not require too much money. It is thus essential to provide for this most basic human need in every rural community. Adequate medical facilities will go a long way towards bringing about a better overall quality of life. In this context Mr. McNamara has observed:[2]

> "The annual income of more than one billion people in the world's poorest countries has risen only by about two dollars (to $150) in the

*Superscript numbers refer to Notes at ends of chapters.

> past decade of massive inflation. But what is beyond the power of any set of statistics to illustrate is the inhuman degradation the vast majority of these individuals are condemned to because of poverty. Malnutrition saps their energy, stunts their bodies and shortens their lives. Illiteracy darkens their minds and forecloses their futures. Simple, preventable diseases maim and kill their children. Squalor and ugliness pollute and poison their surroundings. The miraculous gift of life itself, and all its intrinsic potential—so promising and rewarding to us—is eroded and reduced for them to a desperate effort to survive. The self-perpetuating plight of the absolute poor simply cuts them off from whatever progress there may be in their own societies. Unless specific efforts are made to bring them into the development process, no feasible degree of traditional welfare, or simple redistribution of already inadequate national income, can fundamentally alter the circumstances that impoverish them."

It is a serious blot on the process of science itself to see a large number of people in developing countries suffering from dreadful diseases like leprosy. If coordinated action could enable these countries to get rid of malaria and cholera, there is no doubt that proper effort will help in eradicating other diseases.

Skill-oriented Development

A third vital factor has been experienced by several social service organizations. If attempts are made to bring about improvement in the overall quality of life by pumping in resources, such "Development" is not likely to last long. Quite often, once the supply of resources is removed the process stops altogether. This can be avoided if emphasis is laid on skill-oriented development. Even simple skills are missing in many rural areas. Attempts at development must aim at improving existing skills and training in areas where skills are missing. It is in this context that our programmes are essentially geared towards training activities and non-formal education in areas like tailoring, carpentry, medical assistance, weaving, cane utilization, plastic products, etc. This will enable a rural-based individual to earn his living even after the external resources are pulled out. In fact in our programmes we have found this to be one of the most important human needs—the need to gain a skill and thereby earn a livelihood. No one likes to live on charity; everyone must be encouraged to earn a living by playing a useful role in the community.

A Spiritual Dimension to Development

Another important experience has been that in tackling rural areas one finds a distinct difference between villages which have a spiritual background, and those which do not. Development efforts of religious organizations like Christian missionary organizations or Hindu Temple groups, have had a distinct measure of success. They seem to succeed in inculcating a vital sense of discipline into the community. In areas where such a background is missing development comes slowly, the participation of people is missing and benefits are not lasting. In such cases, help given is demanded as a matter of right and while outside groups work hard to bring about an improvement, idle young people in the villages while away their time in frivolous and unproductive activities. In striking contrast, it is observed that where there is a religious background the village as a whole rises to the occasion and participates because of the common objective before them. It thus appears that more and more religious organizations should be drawn into the process of development. Although many authors have made this point, not much attention seems to be given to it yet. A deliberate attempt must be made to make use of voluntary agencies with a spiritual background in the task of development. E. F. Schumacher[3] says, for example:

> "Economically, our wrong living consists primarily in systematically cultivating greed and envy and thus building up a vast array of totally unwarrantable wants. It is the sin of greed that has delivered us over into the power of the machine. If greed were not the master of modern man—ably assisted by envy—how could it be that the frenzy of economism does not abate as higher standards of living are attained, and that it is precisely the richest societies which pursue their economic advantage with the greatest ruthlessness? The insights of wisdom enable us to see the hollowness and fundamental unsatisfactoriness of a life devoted primarily to the pursuits of material ends, to the neglect of the spiritual; such a life necessarily sets man against man and nation against nation, because man's needs are infinite and infinitude can be achieved only in the spiritual realm, never in the material."

Again, Arnold Toynbee,[4] speaking about spiritual welfare having to take priority over material wealth, writes:

> "Gross national product is not an index of even the economic prosperity of the human beings who compose a nation. Statisticians divide the figure for the GNP by the figure for the national population and call the resulting dividend, average per capita income. This notion is meaningless, and the quantification of it is culpably misleading. A figure for average per capita material damage would be more meaningful, for

although, such areas as housing, the damage from the increases of the GNP in an economically competitive society is unevenly distributed, pollution of the air, soil, water, and other constituents of the natural environment adversely affects all the inhabitants of a country. Pollution may poison a rich mother's child as well as a poor mother's child.

"In a nation in which the first national priority is the increase of the GNP, economic competition between individuals and classes is likely to be intense; therefore, the inequality of the distribution of the GNP is likely to be aggravated. For instance, in present-day Britain, which is partly a welfare state, housing conditions are shockingly bad for the minority below the poverty line. In other words, Britain fails to guarantee the necessities of life to all her citizens.

"I agree that we ought to aim not at gross national product but at gross national welfare. My tests of welfare would be these: The degree of harmony and mutual kindness among the participants in the society; the average *per capita* spiritual welfare, which determines the degree of harmony and mutual kindness, the average standard of self-mastery, which is the key to spiritual welfare, the degree to which the society foregoes profit for the sake of avoiding pollution, both material and spiritual. The last test gauges the extent to which the society has succeeded in giving spiritual welfare priority over material wealth."

Both these authors, and several others, bring an entirely new aspect of development sharply into focus. This may be described as development through contentment or the spiritual dimension to development. One should not entertain the somewhat naive and impractical hope that respect for morality and ethics can be inculcated by preaching—particularly by individuals and institutions who do not themselves practice what they preach. It is heartening to note that UNESCO does mention its concern for the spiritual aspects of human life and culture, though only occasionally, in its publications.[5]

In actual practice, the experience of COSTED has indicated that in communities and villages where the people are imbued with traditions such as fear of sin, faith in God and brotherhood of man, attempts to improve the quality of life evoke a better and more ready response. Such traditions are found deeply rooted amongst the people if they have been derived as part of their culture. They cannot easily be injected afresh as part of their schooling.

Education—an Unseen Important Basic Need

A serious problem in many developing countries is that nearly 50 per cent of the population is illiterate, which itself cuts them off from the process of development. They cannot easily be reached by the methods which are used in other countries. They cannot read popular articles nor can they appreciate the advances of science. One particular problem which is assuming menacing proportions is the problem of population. Forceful methods of controlling population have met with disastrous consequence both politically and socially. The only solution appears to be to educate people about the problems of having a large family. All efforts at development meet with failure because of the population explosion in developing countries. Ironically population growth is high in developing countries where every other form of productivity is low. The utter poverty in which people live is bad enough, worse still is the fact that population growth neutralizes all efforts of development. India has embarked on a crash programme of adult literacy and has set apart vast resources for this programme. It is however, worthwhile to remember that Sri Lanka with near 95 per cent literacy has as many problems as any other developing country, so literacy is not the only thing that should be emphasized.

India's adult literacy programme declares:

> "The main reason why the fruits of development have permeated only to people who were already better off is the illiteracy of India's 230 million adults over the age of 15. Literacy classes are designed to give information on health and family welfare, food and agriculture, handicrafts and rural industries, co-operation, marketing, low economies, sports and cultural activities while imparting the ability to read in the language spoken locally."

In practice it is difficult to achieve this objective unless special methods of reaching illiterate people are evolved. Here again COSTED has evolved a number of non-formal educational packages which are basically audio-visual in approach. Each package consists of slides and a taped commentary (which is translated into the local language of operation) and a number of films are also used. This aspect of "functional literacy" has been one of the successful programmes undertaken by COSTED.

Thus, even while recognizing the advantages of a population limitation, we do not think it easy to tackle the population problem *per se*. It need not be an inescapable prerequisite to development. Substantially the same results can be achieved by a concerted effort to improve the ability to acquire new knowledge (functional literacy). If only we could learn to use the two hands of every person instead of merely complaining about the additional mouth that has to be fed, we could make a happy society.

With the above general comments in mind COSTED itself has initiated minimum needs programmes in different areas of the developing countries. These programmes were conducted in co-operation with local institutions and scientists. The results described below are a summary of some of these programmes.

Satisfying Human Needs—Results of Some Case Studies

In spite of massive financial outlays the problem of rural poverty in India is still very acute. The magnitude of the problem is staggering. The area in which the earlier minimum needs programme has probably had least effect is in providing basic medical amenities in rural areas. The root cause for many rural problems is lack of clean drinking water. In a survey[7] conducted in 1972–3, 152,000 villages were identified as problem villages in respect of the availability of safe drinking water. They were categorized as follows (Table 1).

TABLE 1

	Category	Number of villages
1.	Those which did not have an assured source of drinking water within a reasonable distance (say 1.6 km)	90,656
2.	Those where the source of water supply is susceptible to water borne diseases like cholera and guinea worms	37,041
3.	Those which suffer from excessive salinity, iron or fluorides	24,776
		152,475

The ongoing rural water supply schemes would have covered 95,653 villages with 56,828 yet to be covered, following the last survey the state governments identified some additional villages as problem villages. Taking these villages into account the total number of problem villages to be covered will be 103,000. All these villages are to be provided safe drinking water during 1978–83.

COSTED programmes have concentrated on providing skills and thereby enabling a rural community to aim at integrated development. In certain areas, like medical facilities, external support is required. However, in areas like building a school, or a community centre or a village library, the emphasis has been on providing only material costs but the labour must be provided by the local community. Also, the choice of priority is left to a village coordinating body. Another aspect which received special attention was an in-built evaluation mechanism. For purposes of convenience the programme was divided into five areas and initial statistics were gathered to provide a starting point which can again be compared after a year's effort. Two such examples will be given here.

Among the problem diseases in a village on the rural outskirts of Bangalore were scabies, xarophthalmia, anaemic disorders, hookworm, etc. By concerted medical action these were more or less eradicated. Voluntary medical teams visited the area two days a week and attended to the problems. From a total population of nearly 2000 it was a common sight to see nearly

100–150 patients line up every week with simple preventable diseases. Hence, a special effort was made to prepare audio-visual education packages which were utilized in non-formal education programmes and one could see a change in the living habits of the people. There are other problems like protein/calorie deficiency, etc., which are linked with poverty and for which easy solutions cannot be found. This situation can improve only when the earning capacity of a villager is improved. Special attempts were made to try and improve this. A fairly large percentage of the working population was involved in the profession of weaving. The quality of their fabric was not good and was unable to find a market. This took us into more complex questions like marketing, credit facilities and so on. This once again points out the important fact that science and technology cannot be viewed in isolation and they have a complex interplay with society.[8]

TABLE 2

Staff Position Flow Chart—Rural Technical Work Centre

		Technical Director		
Food and Agricultural Division	Health and Sanitation Division	Housing and Construction Division	Textile Division	Machine Technology Division
Consultant Manager	Consultant Manager	Consultant Manager	Consultant Manager	Consultant Manager
2 Graduates	2 Graduates	2 Graduates	2 Graduates	2 Graduates
5 Intermediate education holders	5 Intermediate education holders	5 Intermediate education holders	5 Intermediate education holders	5 Intermediate education holders
8 High School leaving certificate holders	8 High School leaving certificate holders	8 High School leaving certificate holders	8 High School leaving certificate holders	8 High School leaving certificate holders

Table 2 shows a schematic representation of the work plan that was followed. As can be seen, the main emphasis has been in improving local talent, providing them with an opportunity to find avenues for self-employment and improving their capacity to make a decent living by their own efforts. A striking factor has been the number of unemployed youth. Many of these young able-bodied men turn out to be parasites on society, because they have not learned any skill and for most part of the year (except when they help at harvest time) they remain idle. Utilization of this tremendous human resource will go a long way towards helping developing countries. A conscious effort must be made to set up rural industries. Careful planning will provide a number of ancillary industries which can be set up in rural areas, as support for large urban industries. This was attempted in one section of our programme and proved quite a success. In fact this proved to be the most important basic human need—acquiring of skills. Many people in the villages are able-bodied and should not be allowed to degenerate into social parasites. They certainly do not want to do so but when they find an easy way of getting their daily food by turning into beggars they do so with little hesitation.

Similar programmes are being tried out elsewhere with the support of local governments and private organizations. While this model appears a good blueprint it has not yet been tried in sufficient number of places to be described as a widely applicable model. It does, however, throw up several useful ideas.

(a) Acquisition of skills is an important basic human need. In fact even if a person is provided with his daily food and medical attention he desires to possess some skill and use it to earn his living.

(b) A uniform list of basic human needs cannot be drawn up. Food, clothing, shelter, employment, and basic medical facilities are the primary needs. The level of these needs will grow from time to time and priorities will change. Needs are relative and definitions of minimum standards have little practical significance. In many rural areas people living in loose mud huts (a very primitive dwelling by any standards) want greater attention to be given to providing them with employment and skills than to improving their houses. Of course, a definition of the word "barest minimum" is an open question. The level of tolerance of human misery is an experiment which is not worth performing. People with per capita income of less than $10 a month keep living.

(c) The importance of letting the people in a rural area determine their priorities cannot be over-stressed. The rural people in developing countries are quite frustrated—perhaps justifiably at the slow pace at which "development" trickles to them. Even if the fault of such slow delivery is not with science *per se,* it is necessary for national science programmes to be oriented towards the pursuit of these problems.

(d) Basic human needs are not meant to be doled out as charity. They are the rights of people.[9] "The right to life is the most important of human rights. Life is not just survival but living in health and vigour. Living has no meaning unless it is accompanied by freedom from hunger, diseases, ignorance and poverty. And, for men, life has loftier goals than just living." It is the duty of the country as a whole to provide a meaningful life to its people.

(e) An urgent reorientation of scientific programmes at a national level, and support programmes at international level are called for as Galbraith has observed:[10] "We have probably wasted a good deal of time and effort doing things which were right in themselves but which made little or no contribution to progress because they were done in an environment which was inconsistent with advance."

Research and Basic Human Needs

Science and technology have the potential to provide basic human needs to all mankind. But how can the infrastructure that has been built up be geared to tackle the situation? Admittedly this is not just a question of new scientific research. It is more a question of distribution of resources, improved political commitment and increased social awareness. It is often argued that science and technology are too specialized and are indeed not expected to be utilized for the apparently different task of tackling poverty. We do not agree with this. Poverty and hunger are social problems and must be tackled by the entire society, of which scientists are part. They cannot live in isolation and disclaim all social responsibility. Even if their scientific expertize cannot be directly utilized, their skill and their knowledge can be utilized in the task of nation-building. At times of national crisis all scientists are called upon to tackle priority problems. If leading American scientists had to work at the development of atomic weapons at a crucial time it is not right to expect the scientific community in developing countries to divert all their talents to the task of tackling the problem of providing basic human needs? Some examples will illustrate the potential role which science and technology can play in meeting basic human needs.

The first area in which research and concentrated effort are required is a study into the cultural barriers to better living conditions. Students in every elementary school are taught that they should be clean, should wash themselves every day and so on. Yet in rural areas half the diseases like scabies, hookworm and guinea worm are caused in this very way. Why is it that what they learn is never put into practice? No culture asks them to be unclean, no society asks them to drink obviously polluted or unclean water, or stand in rubbish. Several questions arise. Should we change the method of education, should we use more drastic disciplinary methods when they are young, should we educate their parents about the harm that will come from dirt and the good that derives from cleanliness? Research is called for into this aspect of social change.

Disputes about the definition of basic human needs will last as long as we debate on them. Whichever way we look at it food, clothing, shelter, the right to work and the right to learn appear as prominent basic needs. It is the relative priority that may vary from time to time and from place to place. No society wants to live without electricity if it has an opportunity, but if its people do not have water to drink and food to eat, electricity has little meaning for them. Similarly, for people living in rural areas, in acute poverty, food is more important than a well-planned house.

Let us look at other areas where research is called for. Table 3 shows the availability of food in different sectors. It shows the low per capita food consumption in developing countries. Concentrated work on agriculture will

TABLE 3

Availability of Food

	World	Developed	Developing
Population			
millions	4013	1057	2956
Cereal food production, millions of tons	1233	681	552
Available food, kilograms/year/person	307	644	187
calories/day/person	4205	8825	2562
Food consumption, total grams/day/person	725	2000	400
calories as grains/day/person	3625	12000	2300
calories/day/person	2263	3000	2000

Source: Data from *UN Statistical Yearbook.*

definitely produce more benefits. If technology is available one cannot easily see why only developing countries should continue to have these problems.

Over the past 25 years, grain production has kept pace with the rising world population. There have been year-to-year fluctuations, of course, but the maximum annual fluctuation from the mean never exceeded 60 million tons, about 5 per cent. This record of increasing food production has been due largely to continuous increase in the agricultural efficiency of the food-exporting countries. Smaller increases must be expected in the future because the major inputs, the hybrid grains introduced in the 1960s, the fertilizers and the pesticides are all now approaching optimum general use. Furthermore, most of the high quality farmland is now in use. Yields will continue to increase in the food-exporting countries, but logistic, economic and social arguments combine to dictate that most of the doubling of food-grain production which will be needed in the next 25 years must be obtained from crops grown in the poor countries. It is of crucial importance that a global effort be made to apply and adapt the best existing agricultural science and technology to increase food production in the countries in which the major population increases will occur.

Technology does provide means for improving agricultural production where it is needed. New high yielding tropical varieties, proven irrigation technologies, faster-growing crops which reduce weather hazards and which respond well to fertilizers are examples. The obstacles to their application are mainly political, social and economic. Until both the poor countries and the countries which provide aid give genuine top priority to developing rural areas and improving transport, water supply and amenities for the villages, it will be impossible to feed the growing population and we cannot foresee other than a food crisis of major proportions in many areas.[11]

Another important area is that of providing employment to all people. In its report to the World Employment Conference in Geneva, the International Labour Organization computes that between now and the year 2001, employment will have to be created for 1 billion workers in the developing countries.

Within these aggregates, the unemployment of educated people in the developing countries which was around 9 million in 1970 is now around 15 million, India alone registering a total of 5 million job seekers. There is in addition the problem of underemployment which is estimated to cover 30 per cent of the working population in these countries. In the absence of a system of social security, really no one in a developing country can afford to be completely unemployed, unless he is living off a relative as a member, for example, of a joint family. Everyone has to work at something, even it be at literally scratching at a dry barren ground for a living. There are conceptual problems in defining unemployment and underemployment which are the subject of perennial debate among economists, the broad meaning of underemployment being that a person is working far less than the time period that he can. There is also the problem of disguised unemployment wherein a proportion (varying from 10 to 20) of those working on agriculture to fill in time can be removed from that sector without any decline in agricultural output.

In this somewhat daunting employment profile, four lines of action are suggested for the developing countries. First, the development strategy which in the sixties and even today has been aimed at GNP growth has to be redirected towards a mix of employment generation and growth. Its aim should be the increased production of goods of mass consumption.

Second, there is a special task for science and technology in these countries. Today science and technological institutions are alienated from the agricultural and manufacturing industries and have become consumption items instead of production inputs and investment items. In this context, what is needed is for the science and technology of the developing countries to develop the technologies needed by them for meeting, on the one hand, their special and large manpower resources and, on the other hand, and in the process, become the infrastructure for the full and improved production of goods and services needed by the poor masses in agriculture, animal husbandry, fisheries, edible oils, cereals, cloth, education, health and nutrition, water supply, housing and small and cottage industries which today have little science and technology back-up.

Speaking about these problems Mr. T. A. Pai, Union Minister for Industries and Civil Supplies said:[13]

> "Our industry serves only about 60 million people or about 10 per cent of the total population. Like our medical graduates, it is foreign-oriented and city-dominated." He goes on: "We have perhaps all the leading drug producers of the world operating in India, yet we have one of the highest mortality rates. The organized industry and particularly the foreign companies concentrate on producing high priced drugs, tonics and some remedies of dubious curative values. Drugs for common ailments are

either not made in sufficient quantity or are so highly priced as to be out of reach of the common man."

The problem of meeting basic human needs is not merely one of resources. Even where massive inputs are provided the results are poor. We believe it necessary, in fact essential, for some international organization to undertake the task of taking a few rural areas and building them up as model areas where basic needs can be met through the proper application of science and technology. To some extent this has been the objective of COSTED Rural Development programme and a good measure of success has been achieved. This has to be repeated in several select areas to test the validity of the conclusions arrived at. Such an attempt will bring immense credit to the international scientific community and will serve to demonstrate the practical contributions which science and technology can make at grass root levels.

References

1. *Science and Technology for Integrated Rural Development,* p. 8, COSTED Seminar Proceedings, Madras, 1977.
2. Annual Meeting of the World Bank, Manila.
3. *Small is Beautiful,* by E. F. Schumacher.
4. *Choose Life,* by Arnold Toynbee and Diasaku Ikeda.
5. *Thinking Ahead,* UNESCO Publication 1977.
6. *Yojana,* October 1978, Indian Planning Commission Document.
7. *Yojana,* page 21, 15 August 1978.
8. S. Radhakrishna, *Science and Technology for Integrated Rural Development.*
9. ASHA document, National Paper of a Typical Developing Country (ASHA), p. 37.
10. J. K. Galbraith, *Economic Development in Perspective,* Faucett Publishers, N.Y., 1962, pp. 12–15.
11. *Science, A Resource for Human Kind,* Bellagio Conference, 1976.
12. *Science in the Service of Basic Human Needs,* pages 15 and 16.
13. *Science in the Service of Basic Human Needs,* COSTED Seminar Proceedings, Hyderabad, India, 1976.

SECTION 4

Conclusions

Conclusions

SAM NILSSON

From the contents of this book it is clear that much has been achieved by the UNESCO programme on Research and Human Needs since its inception in 1975–6. With only modest financial resources, scientists in developed as well as developing countries have been stimulated to initiate activities and research which would otherwise probably not have been carried out. It is fair to say that the programme has poineered a major international research effort on the definition and understanding of human needs and wants. This is the more significant because it has been done during a period when an intense debate has been going on within the UN system about the meaning of "Basic Human Needs" strategy, which was launched by the World Bank and ILO about 1976. The research and human needs programme has also begun to build up a system for providing information on the problems of human needs and multidisciplinary research related to such problems.

Most important of all, however, are the local initiatives stimulated by UNESCO. In Latin America, Africa and Asia the first promising steps have been taken towards self-perpetuating programmes which will include both theoretical and action-oriented activities. This is in accord with the contribution to this book by Soedjatmoko, who argues strongly for the increasing consideration of local socio-cultural values in the development process. The emphasis on local activities also supports the arguments presented by Hogbe-Nlend who urges the scientific community in developing countries to re-orient its research towards the agricultural, nutritional, medical, technical and cultural needs of their own countries rather than pursuing the research goals of developed countries. It is gratifying to observe that many of the actions suggested by Udgaonkar are being implemented within the programme or by other organizations. It is also time, of course, that many scientists are working on the conceptual and methodological problems of the programme in Europe and the Soviet Union. It is important that scientists from both developing and developed countries can work together on these problems which are, in fact, of equal significance in both societies.

After this promising start, the time has come to consolidate the programme and define its objectives more clearly, especially in view of other programmes and the medium term plan of UNESCO. The research and human needs programme is meant to be a supplement to the normal priorities and values of basic research confined to strictly disciplinary subjects. The programme gives the research community an opportunity to be more relevant in attacking the problems of satisfying human needs. The programme is thus concerned with advancing communication lines other than the normal ones within the scientific community, for example, between researcher and problem area; between researchers in different fields including social sciences; and between researchers on the one hand and policy makers and the general public on the other.

The main objectives of the programme can be summarized as follows: to mobilize and re-orient scientific research towards the problems of human needs, defined as clearly as possible, and to promote interaction between the various interest groups mentioned above.

In December 1980 UNESCO called together a group of scientists from developing and developed countries to define the structure and content of the programme for the next few years. It recommended that the programme be divided in three main parts: (a) research on the formulation of needs; (b) research on the fulfilment of human needs, through information and stimulation of new research on the problems, and through information on possible solutions; and (c) interdisciplinary pilot projects.

The re-orientation of research and mobilization of the world scientific community towards new priorities will take a long time. The research and human needs programme, however, was started at the right time, when the goals of development and the relevance of science and technology were beginning to be quesioned. Today a new awareness is emerging about the nature and magnitude of the world's problems and about how those problems relate to different countries. The programme will therefore be able to play a more significant role during the next few years than it has so far, and it is with this in mind that we now turn some reflections on the future.

Decades of Transition

For the first time in history man has the scientific and technological capacity to eliminate poverty, starvation, debilitating diseases and water and housing problems for everybody, It is not enough, however, to organize an all-out mobilization of scientific and technological potential even if this were practically possible. Major institutional changes are needed in the private and public sectors, or the changes will be too slow. Attitudes towards science, technology and society must be modified and we must try to understand the problems more clearly before we jump to conclusions about the solutions.

In a recent paper[1] prepared by IFIAS for the Population Division of UNESCO it is pointed out that present world trends and changes are so rapid, so interactive and so widespread that mere extrapolation from today can give a quite misleading picture of the nature of future developments. The main reason for this is the existence of unexpected interactions between policies shaped in isolation. Sometimes these interactions reinforce each other, sometimes they create unexpected conflicts and contradictions. Looking at some of the most important trends it is hard to avoid the conclusion that we are entering a period of transition which may last 30–50 years. Whether this transition will lead to a new type of society is impossible to tell, but it is likely that both the technical structure and the value basis of world societies will change. The transition will be forced upon society by a whole range of difficult causes, but three appear to be dominant.

The first is the large increase in the total population of the world. This demographic explosion will make heavy demands on the world's resources and environment. It will demand an integrated approach to resource management and will necessitate government of high quality if a life of dignity and decent prosperity is to be provided for the new inhabitants. The availability of sufficient food is the first need of all societies, and it should be possible technically to provide enough for the new multitudes. The problem of its equitable **distribution,** however, is likely to prove much more difficult. The difficulties are essentially political, economic and logistic, and it is unlikely that they can be resolved without a great deal of international co-operation and an economic system which goes beyond those in operation today. The technical problems, although less intractable, have to be seen in terms of improved agricultural research, land availability, land tenure systems, the need to avoid short-term solutions which are difficult to sustain, problems of water availability and, above all, energy constraints. It has taken the industrialized countries about 150 years to approach zero population growth. If we cannot learn from this experience and reach it quicker in the developing countries we will face overwhelming problems during the next few decades.

The second factor is the need for a complete rebuilding of world energy supplies as petroleum becomes scarcer and more costly.[2] This change is taking place at the same time as population growth and both will have very high demands for capital. The new energy system will comprise a wider range of energy resources with an initially gradual, but later accelerating trend towards the use of renewable resources. The big question is whether research development and construction can be completed before major shortages and the high cost of traditional fuels bring serious economic and social difficulties.

The third factor during the transition period is likely to be the impact of new technologies, especially those based on microelectronics and on progress in biological research. Electronics will virtually revolutionize industry—its processes, demands for materials and energy and its international distribution—with great possibilities for everybody. But it has also deep implications for social change. Information and communications are undergoing a complete transformation, and the automation of industry, of office work and of other tertiary sector activities will give rise to great transitional difficulties of unemployment and creation of new types of jobs.[3] These problems can be solved only by careful planning well in advance and new concepts and attitudes to replace those associated today with work, employment, leisure and unemployment.

The second technical change, based on biology, may in the long run favour the tropical countries because this change could revolutionize the use of biomass, making contributions both to the manufacture of food and ultimately, perhaps, energy production. Eventually such countries could become major exporters of energy and food.

These three changes—population increase, energy renewal and the products of the new technologies—thus create great problems of transition but also hold out great hopes for future prosperity. All three will make enormous demands on capital, and will necessitate profound social and cultural adaptations and mutual understanding. For the developing countries, which face problems of food production and infrastructure building at the same time as the need to reassess development objectives, the difficulties will be considerable. The trends outlined here will increase the interdependence of nations and help bring about new groupings of countries and new alliances, new methods of international co-operation and new economic arrangements.

The problems of the transition period and the need to establish a number of "holding" positions should make it evident to all countries that there is a need for some sort of New International Economic Order as a matter of common self-interest. This is emphasized in the contribution by Soedjatmoko and was a leading theme also in the report of the Brandt Commission.[4] All countries will, in fact, be developing countries during this transition to new types of society. However important the problems of equity between the nations may be, and however vital the economic questions, the main problems are likely to be political, social and human, and in planning for the New International Order, a broad approach will have to be taken so that cultural diversity can be preserved and strengthened within a harmonious framework in which the fundamental human needs are provided and equitably distributed. In preparation for the transition, each country should try to determine its "carrying capacity", which ought to be seen not as a figure fixed by solely material constraints, but as a range of alternative figures within a range governed by the socio-cultural options. In other words each country should determine as precisely as possible its carrying capacity in view of its population status, economic development prospects, education, technology, and environmental conditions.

It is in this context the research and human needs programme of UNESCO should be seen. UNESCO's role should be to stimulate the scientific community to become more actively involved in understanding the problems and in developing solutions which make the transitions and adaptations as smooth as possible. Although the problems will become more global and interdependent, most solutions will have to be articulated in national terms. The solutions—technical or economic—should not be imposed from outside on any society. Danzin makes an interesting analogy between the evolution of society and that of biochemical systems. He suggests that at a certain point any society may make an irreversible shift and proceed in either of a few and essentially unpredictable directions. This means that social changes may not always proceed smoothly, but in a series of abrupt changes, which makes adaptation difficult. The present world situation, characterized as it is by simultaneous high unemployment and high inflation in the industrialized countries at a time when the developing countries need help and co-operation more than ever, may well be considered to be such a critical moment of change. As Danzin points out, this offers several solutions of which some may not be at all obvious. In such a state of transition individual societies and the world as a whole might benefit from analyses of what the likely alternatives are and of the consequences of adaptation, in order to be better prepared for the inevitable changes. The scientific community has a special responsibility in this area and UNESCO can, through its research and human needs programme, assist member states with multidisciplinary analyses of vulnerability, resilience, and carrying capacity and with the utilization of science and technology to remedy the worst effects of the transition.

The Premises of Contemporary Society

Many recent reports and conferences bear witness to an increasing awareness in political quarters and in the intellectual community about the rapidly changing world situation. This is, of course, encouraging and indeed vital if practical programmes are to be introduced. The potential of science and technology for leading an attack on development problems is often mentioned. Many believe that we are on the threshold of a new industrial revolution, brought about by microelectronics and biotechnology, which will constitute the basis of the continued modernization and industrialization of more and more countries. Production and consumption will continue as before, only in greater and greater quantity. Others believe that new life styles and a new consciousness, characterized by a greater understanding of man's place in the ecosystem and the role of spiritual and humanistic values, will begin to replace the materialistic values that now prevail. In this connection it is interesting to reflect for a moment on the intellectual basis of the last Industrial Revolution. The "mechanical world view" originated with just three men—Francis Bacon, René Descartes and Isaac Newton. Three hundred years later we are still living off their ideas. The world view as advanced by these three thinkers dealt exclusively with material in motion, because that was the only thing that could be mathematically measured. It was a view made for machines, not people, and the mechanical paradigm proved to be irresistible. It was simple, it was predictable and, above all, it worked. The erratic behaviour of people and the imperfect working of governments and the economy did not always seem to square with this view of the world, but this dilemma could be quickly resolved. If society was misbehaving then it could only be because it was not adhering to the neutral laws of nature.

Two men immediately set out to discover the relationship between these universal laws and the workings of society. John Locke brought the workings of government and society into line with the mechanical paradigm, and Adam Smith did the same for the economy. "The negation of Nature is the way to happiness," Locke declared. It is no exaggeration to say that Locke formulated the basic philosophy on which the industrial world is founded; that man is born a blank slate, *Tabula Rasa,* on which anything can be written. People must be "effectively emancipated from the bonds of nature", was his conviction. Smith, like Locke, believed that the basis of all human activity is material self-interest. Any notion of morality in economics was effectively removed by the "invisible hand" which has been elevated to a religion by free market economists. The concept of the "survival of the fittest" which emerged in Darwin's Theory of Evolution was a further support for this mechanical world view, showing as it appeared to that self-interest leads to increased well-being, which leads to increased order in nature. (In physical terms this means decreased disorder, or decreased entropy, according to the Second Law of Thermodynamics.) The logical conclusions of these grand paradigms based on philosophical and scientific achievements was: "The more material well-being we amass, the more ordered will the world become." Progress then became a process of amassing greater and greater material abundance, which was assumed to result in an ever more ordered world, both physically and socially. Science and technology are the tools for getting the job done, and this is why scientific research and its applications are considered by so many to represent the most important force of change and development.

Today we know that this view is too limited, even if most economists still seem to believe that the more energy in terms of oil or human labour we add to nature's resources the greater will be the value added. Everything in nature is, according to Locke, considered to be waste until human labour or technology has been added to it, transforming it to something that can be exchanged and consumed for the good of society. By turning the laws of thermodynamics upside-down modern economic theory has completely misinterpreted the entire basis of all economic activity. Whenever energy is extracted from the natural environment and processed by society, part of it is always dissipated or wasted at every stage, until all of it, including that which is made into products for consumption, ends up in one form or another as waste at the end of the line. Most economists simply cannot accept this simple truth.

We are surrounded by manifestations of the failure of the mechanical world paradigm. For example, to produce one can of corn containing 270 calories, the average US farmer uses up 2790 calories, most of it as fuel to run the farm machinery and as synthetic fertilizers and pesticides. The food we eat today is more dependent on oil than on soil. Another example is in transportation. It takes only half the energy to move one passenger 1 mile by public transit as it does by private car. It takes only 670 BTUs (British Thermal Units) of energy to ship 1 ton of freight 1 mile by train and over 2800 BTUs by truck. The automobile has become a central feature of our fossil fuel culture. The massive physical and social disorders caused by the car in the form of roads, urban expansion, pollution, accidents and stress, constitute a discouraging example of what happens when an economic system fails to take into account the effects of the entropy law until too late. The benefits we have derived from the car—and they are certainly many—must now be seen in the light of the even greater penalties we are forced to pay. The total bill is probably more than we can afford.

Another phenomenon of modernization is structural centralization. Industries become bigger and bigger and more and more centralized. We have had cities for many thousand years but the old cities, with the exception of Rome during the Roman Empire, were all self-sufficient in food, energy and most services. With the Industrial Revolution this changed. London became the first city with more than 1 million people. By 1950 there were 75 cities of at least 1 million, and if today's projections prevail, more than half the world's population (more than 3 billion) will be living in cities by the year 2000. This will probably represent the biggest mass migration in history.

A typical urban area of 1 million people requires a daily input of about 2000 tons of food which is entirely dependent on a fuel-driven agriculture and transportation system. The same urban area also requires for its sustenance a daily input of 9500 tons of fuel and 650,000 tons of fresh water. Like the Roman Empire, industrialized societies become more and more dependent for their survival on resources far away from their cities, or even from their national boundaries. We know what eventually happened to the Roman Empire.

The Need for a New Development Approach

Dr. Ernest Gellner of the London School of Economics is convinced that the industrialization–modernization process will continue simply because the alternatives are less attractive. Few societies have been more oppressive to the individual than the agrarian, because there were always food shortages and this granted to a few direct control over the survival of many. Thanks to the scientific and technological revolution in the industrialized societies, survival is no longer precarious, and wealth no longer buys power through the control of lives. On the whole, individual freedom and the quality of life have improved enormously during the past couple of decades. Never before in history have so many people in the world had abundant food, health, education, housing, and employment, but at the same time we have to admit that never before have so many people been deprived of precisely these things. Even if scientific research and technology on their own cannot guarantee a decent quality of life for everybody, we must improve the distribution of science and technology and the education connected with it.

There is a need for new development approaches not only in the developing countries but also in the developed countries. We must begin to search for approaches appropriate to both and for solutions which yield benefits to as many as possible and negative effects to none, if possible — 'positive sum-game' solutions, as they are called in game theory. The values guiding us today, the available technological tools, and the institutional mechanisms were made for a simpler and a smaller world. Below are given some desiderata which we consider essential for a new development approach and for the research and human needs programme.

Every human being should have the right to a life that makes possible self-realization, both spiritually and materially.

Because society consists of many individuals, self-realization should be co-operative rather than competitive.

Problems and solutions should be viewed in a "holistic" manner, so as to take into account as many as possible of the interacting factors.

Development programmes should respect the fragility and diversity of the ecosystem and respect local social and cultural values.

Global development should strive for harmony and mutual respect among different cultures and beliefs.

A system of global taxation of arms expenditure should be developed, and the money raised used for the financing of international research on human needs problems.

Most of these requirements have already been taken into account in the research and human needs programme by UNESCO. But much stronger efforts must be made in the future to refine and promote development approaches which help individual countries to utilize the global pool of knowledge for their own development. The scientific community has a great responsibility in this effort, and we need to create practical and flexible mechanisms which attract rather than discourage scientists and other intellectuals from taking their share of the challenges. Below are a few suggestions about possible new approaches to development in general and the role of science and technology in particular.

Each country should try to articulate its development goals in the light of prevailing global trends. It is particularly important that the time-perspective should be sufficiently long, at least 5–10 years.

Methods should be developed for integrating policies for scientific research into the national development policy of developing countries.

There is no science policy, however, that is universally applicable to all countries. Each country must decide on its goals and the manner by which it wants to satisfy them. One approach would be the assessment of the "carrying capacity" of a country. This has been defined as "the number of persons sharing a given territory, who can for the foreseeable future, sustain a given physical standard of living, utilizing energy, land, water, skills and organisation".

Assessments of carrying capacity give a first impression of the vulnerability or resilience of a country and thus indicate what the priorities in research should be, in relation to human needs and social goals. This method of coupling research and human needs was first suggested in the biennial report submitted to the research and human needs programme by IFIAS in 1978. The carrying capacity methodology needs further refinement; in particular it is difficult to include socio-cultural parameters. UNESCO can play a very important role in this, and in the introduction of the technique for policy purposes in developing countries.

Policy-makers do not always realize that scientific research takes a long time to yield practical results. This is why it is so important to integrate science policy organically into the general development policy of the country. There is always a risk that fundamental research will suffer from an over-utilitarian approach and some research must be carried out simply for its own cultural value. But not all the research must be done within the country itself, which is why two other points should be emphasized.

"Twin" arrangements should be established between research institutions in industrial and developing countries. A few excellent examples already exist in the research and human needs programme and should be further promoted by UNESCO.

The second point is the importance of innovation. Innovations are not usually made by scientists but by entrepreneurs. Developing countries should foster a social climate that stimulates innovation, and makes it part of the national development programme.

This discussion has attempted to suggest a possible future for the research and human needs programme in the light of world problems. A few specific suggestions have been made about the ingredients which are likely to be particularly important, UNESCO, with the broadest mandate of all UN organizations, should do its utmost to promote and support research that improves our understanding of the problems in these areas, and to assist in a major mobilization of research and technology for the development of solutions to human needs problems. The process will be long and difficult and will need substantial financial resources. But eventually more and more member states will be able to attain a state of self-reliance which will enable them to utilize the interdependence of nations rather than remaining its victims.

Notes

1. *Population, Resources and Society—the Need for Integrative Policies,* Report by IFIAS to UNESCO, October 1980.
2. *A Global Energy Supply Model,* M. Slesser *et al.*, 1980.
3. *The Impact of Microelectronics on Employment in Western Europe in the 1980s,* European Trade Union Institute, Brussels, 1980.
4. *North-South, A Programme for Survival,* Pan Books, London, 1980.

APPENDIX
An Opportunity in Interdependence: The Fulfilment of Basic Needs

MAGDA CORDELL McHALE*

The 1977 Houston Declaration[1] reminds us that:

> "The central purpose of (all) economic and social development is to meet human needs . . ."
> ". . . The satisfaction of human needs is indeed the whole purpose of growth, trade and investment, development assistance, the world food system, population policy, energy planning, commodity stabilization, ocean management, environment protection, monetary reform, and of arms control."

There is a growing realization that the central purpose of all development and growth, whether economic, cultural or political, is to meet human needs—and that the most urgent challenge is to deal with the persistent mass poverty around the world whose continued existence is an affront to human dignity.

Attention to basic needs, in terms of food supply, education, health and housing, has always been a strong component in traditional development and aid programmes. But success or failure in development has always been measured by examining productivity, the growth of GNP, and maximal resource exploitation. Today it is recognized that success requires a much more complex integration of social, cultural, economic and environmental factors, within which the needs of the individual citizen for an adequate standard and quality of life are *the* basic measure.

But the attempt to eliminate mass poverty by basic needs strategies is a new approach which turns the development process on its head. Rather than simply being "more of the same", the *integrated* approach to the fulfilment of basic human needs represents a new paradigm for the development process and a touchstone for overall development itself.

---

* Magda Cordell McHale is Director of the Center for Integrative Studies at the State University of New York at Buffalo, NY 14214.

Basic Needs of People

The basic needs approach is a natural development of the idea that all classes in a nation have equal claim to what is essential for a decent standard of life to a recognition that all people, regardless of nationality, have that same claim. The idea that people might reasonably expect their needs for adequate food, health, education and shelter to be met as basic rights is now more than just a question of morals: it is a self-evident proposition, in the interest of not only the more affluent nations but of the world community as a whole.

Most of our global instabilities, uncertainties and frustrations are the result of social tensions, economic disruptions, and political pressures which arise from unjust conditions around the world in all societies. Visible manifestations are the increase in social turbulence and unrest, not new in themselves, but unique to our period in their extent and spread throughout all societies.

Shared Problems

We face two different and coexistent realities. One is the reality of regular meals, education, employment, entertainment and margins of safety against the unexpected. The other is compounded from hunger, disease, ignorance, insecurity, daily uncertainty and fear. This second reality is experienced by almost two-thirds of the world's population, and exists in all societies. If it goes uncorrected, this reality may swiftly engulf us all.

The traditional solutions to such problems are dangerously ineffective. We are living in an interdependent world, and are all part of a global enterprise where our very survival depends on each other. The issues of food supplies, epidemic control, growth or decline of population, environmental conservation or deterioration, development and employment are intertwined in ways that it would have been difficult to conceive of, let alone predict, several decades ago. All our local issues of unrest and dissatisfaction are symptoms of these shared global problems on an individual, national or regional level. We need the willingness to admit that the solutions to our "shared world problems" do not lie in mutual distrust, but rather in mutual support to combat the trends that lead to inequities and injustices.

The Interdependence of Basic Needs

Inherent in the basic needs approach is the concept of interdependence. Basic human requirements interrelate in at least three ways:

(1) The satisfaction of one need may depend on a more pressing one being satisfied, first, for example, hunger and health are related in this way (functional interdependence).

(2) A single person or unit is rarely able to cope with needs in isolation, but depends on the support and services of society as a whole (systemic interdependence).
(3) The "quality of life" requirements depend on the availability of alternatives, options and choices accessible to the individual.

To look at one area of need at a time is unlikely to break this cycle of deprivation. A long and healthy life, for example, is not determined by health services alone but by the combination of medical facilities, food, clothing, housing and fuel, and by such other factors as meaningful work, education, recreation and security, by the satisfaction of values and by the desire for living itself.

There are more than enough resources and capacities available to eradicate poverty within a relatively short time, despite the current cries about intrinsic shortages of food, materials and money. The obstacles are not material but human. They are principally those of our conventional economic, political and institutional arrangements, which on the whole perpetuate poverty, inequity and injustices while squandering a major part of the world's wealth on armaments and other forms of national aggrandisement. There is, however, no one obvious way to meet everyone's needs. Each country has different problems and different local capacities; each country must therefore choose its own path within a framework of collective self-reliance and regional and international co-operation.

Opportunities in Interdependence

In the past few years, the harsh realities of a fragile global interdependence affecting food, fuel, materials and jobs have forced even the advanced nations to reconsider the direction of their own development. It is suddenly borne upon even the more fortunate that we are all in the same leaky boat. The rich cannot stay rich unless the poor get a bit richer, too; spreading the wealth around means eventually that everybody gets more, not less. It has taken the Western nations over half a century to realize this simple truth, which is the heart of their own development.

The recent past has taught us that we can no longer remain strangers to each other. Our previously separate local community has now become a completely interdependent global community. Most of our production, distribution, transportation and communications systems now require and are dependent upon the resources of the entire earth for the metals and minerals from which they are built and for the energy to run them. The preservation of the global environment is a joint responsibility in all our interests, and the key to survival for us all.

Though it has often been pointed out that many of the world's problems are most clearly evident in the developing countries, it is evident that all countries are undergoing a painful transition into "modernity". We all face severe dislocations, and the obsolescence of our socio-economic and political structures. The trends that bring us collectively into a planetary community are not those visible bonds between nations but rather the fact that many of our activities are overarching forces—social, institutional, technological—whose effects flow across national boundaries (like acid rain) and whose maintenance transcends, and transforms the concept of local national sovereignty into a transnational character.

We need each other for access to markets, knowledge and techniques. At the world economy level, the picture is of growing interdependence not only in relation to markets and raw materials but of increased financial interpenetration of local economies by international investments. The world economy is no longer a structure of competing national interests, but rather a global trading post.

We have learned that the world that has been made has not come about as a result of any political or ideological notions—but by the operation of social, economic, scientific and technological facts. It is these less visible forces, which do not follow the lines of political maps, that have changed the world. Our reluctance to react to these changes lies in our historical and cultural inheritance. Bad memories of past experiences must not divide us as people, nor influence our behaviour towards each other. A respect for cultural diversity and past historical circumstances has to coexist with our present need for one another. As the Brandt Commission Report puts it:

> Focusing on questions of historical guilt will not provide answers to the crucial problem of self-responsibility on which alone mutual respect can build. Self-righteousness will neither create jobs nor feed hungry mouths.[2]

Different individual world views can coexist and enrich the cultural and social environment. Although our various ideological viewpoints may differ considerably, the same social needs and expectations are experienced everywhere. We have to put our interests in the context of mutual benefits, and the satisfaction of basic needs can provide the unifying framework for acquiring the benefits that are inherent in interdependence. The challenge is to realize the potential of each individual through social, economic, environmental and technological innovation and development within a global interdependence.

References

[1] Declaration of the International Conference on Human Needs, University of Houston, 1977.

[2] *North-South: A Programme for Survival,* The Brandt Commission Report (MIT Press, Cambridge, 1980, pp. 24-25).